Introduction

During the coming decades, global change will impact food and water security in significant but highly uncertain ways. There are strong indications that developing countries will bear the brunt of the consequences, particularly from climate change. In Sub-Saharan Africa, agriculture—the mainstay of rural livelihoods—is particularly vulnerable to the adverse impacts of climate change, and the adaptive capacity of rural smallholders is extremely low. Consequently, it is important to understand the impacts of global change on agriculture and natural resources in Sub-Saharan Africa and to identify informed and effective adaptation measures and investment priorities to alleviate the harmful impacts of global change.

This set of briefs assesses these issues with a focus on the Nile Basin in Ethiopia and the Limpopo Basin in South Africa. Authors identify climate change impacts on agricultural productivity and food production; assess the vulnerability of the farming sector and farm households to climate variability and change; examine climate change perceptions; and suggest adaptation strategies at the farm, basin, and national levels alongside the associated investments needed to implement such strategies.

We are grateful to the authors for their research and analyses, to the reviewers for their constructive comments, and to Mary Jane Banks and Ashley St. Thomas for their editorial assistance. We gratefully acknowledge financial support from the Federal Ministry for Economic Cooperation and Development, Germany. We hope that the insights on climate change impacts and adaptation options presented here will contribute to policy changes that profoundly increase the capacity of the rural poor in Sub-Saharan Africa to adapt to climate change.

Claudia Ringler
Senior Research Fellow, International Food Policy Research Institute, Washington, DC

Elizabeth Bryan
Research Analyst, International Food Policy Research Institute, Washington, DC

Rashid M. Hassan
Director, Centre for Environmental Economics and Policy in Africa, University of Pretoria, South Africa

Tekie Alemu
Dean, School of Economics, Addis Ababa University, Ethiopia

Marya Hillesland
PhD candidate, American University, Washington, DC

IFPRI®

International Food Policy Research Institute
2033 K Street, NW
Washington, DC 20006-1002 USA
Phone: +1-202-862-5600
Fax: +1-202-467-4439
Email: ifpri@cgiar.org
Skype: ifprihomeoffice

www.ifpri.org

INTERNATIONAL FOOD
POLICY RESEARCH INSTITUTE
sustainable solutions for ending hunger and poverty
IFPRI®
Supported by the CGIAR

IFPRI Research Brief **15–1**

HOW CAN AFRICAN AGRICULTURE ADAPT TO CLIMATE CHANGE? INSIGHTS FROM ETHIOPIA AND SOUTH AFRICA

The Impact of Climate Variability and Climate Change on Water and Food Outcomes

A Framework for Analysis

Claudia Ringler

Over the coming decades, global change will have an impact on food and water security in significant and highly uncertain ways, and there are strong indications that developing countries will bear the brunt of the adverse consequences, particularly from climate change. This is largely because poverty levels are high, and developing-country capacity to adapt to global change is weak. Furthermore, the rural populations of developing countries—for whom agricultural production is the primary source of direct and indirect employment and income—will be most affected due agriculture's vulnerability to global change processes. The agricultural sector is the largest consumer of water resources, and variability in water supply has a major influence on health and welfare in poor areas. With water scarcity and extreme weather events expected to increase under climate change, water security could decline significantly in rural areas. Consequently, it is important to understand the impacts of global change (in terms of climate, demography, technology, and so on) on agriculture and natural resources in developing countries and to develop adaptive capacity to respond to these impacts. Moreover, there is a need to develop informed and effective adaptation measures and investment options that can be taken now to alleviate adverse impacts of global change in the future.

FRAMEWORK FOR ANALYSIS

While food and water security are largely determined by actions taken at the local or national levels, global factors—such as world food trade, global climate and climate change, and competition for water—also affect food and water security locally. Moreover, human alteration of land use patterns, urbanization, elimination of wetlands, nutrient overloading in water systems, and other biophysical changes could dramatically affect the ability of the global water cycle to support needed food production. The development of policies that mitigate adverse impacts, enhance positive impacts, and support adaptation to climate and global change, together with enhancing local food and water security, therefore requires an understanding of the interactions among local, basin-level, national, and global factors.

Thus, analysis of strategies for increased food and water security must take into account relevant hydrologic, agronomic, economic, social, and environmental processes at global, regional, national, basin, and local levels (Figure 1). This could be done following the paradigm of "strategic cyclical scaling" devised by Root and Schneider (see further reading), which incorporates large- and small-scale research studies to improve our understanding of complex environmental systems and allow more reliable projections of the ecological, economic, and social consequences of global change. Process-based, bottom-up relationships are used to predict larger scale behavior, which is then tested against large-scale data for a top-down evaluation. Cycling between large and small scales should thus produce more credible overall results.

Figure 1 Global Change, Spatial Scales, and Adaptation Strategies

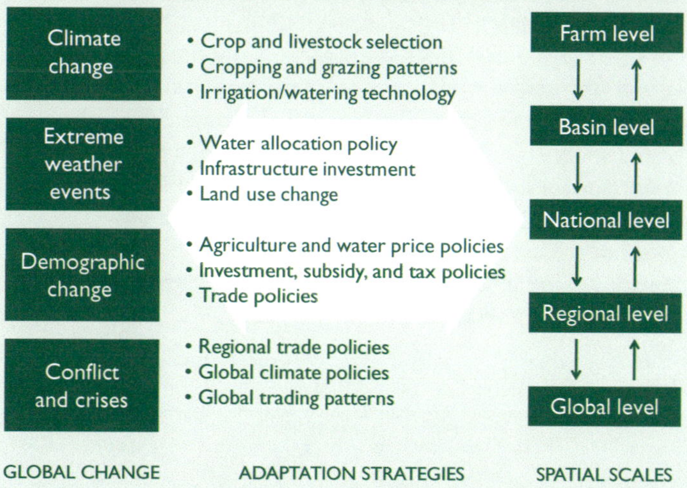

SOURCE: C. Ringler, The impact of climate variability and climate change on water and food outcomes: A framework for analysis, in C. van Bers, D. Petry, and C. Pahl-Wostl, eds., *Global assessments: Bridging scales and linking to policy*, GWSP Issues in Global Water System Research No. 2, http://www.gwsp.org/downloads/gwsp_issues_no2.pdf, 2007.

RESEARCH ACTIVITIES

A project supported by Germany's Federal Ministry for Economic Cooperation and Development, entitled "Food and Water Security under Global Change: Developing Adaptive Capacity with a Focus on Rural Africa," has conducted research on adaptation to climate change at various scales. This project, which is associated with the Challenge Program on Water and Food under the Consultative Group on International Agricultural Research (CGIAR), involved close collaboration with researchers at the Center for Environmental Economics and Policy in Africa, the Ethiopian Development Research Institute, the Ethiopian Economics Association, and the University of Hamburg.

At the local level, farm household surveys were implemented in the Nile River Basin of Ethiopia and the Limpopo River Basin of South Africa to examine vulnerability to shocks, perceptions of long-term changes in climate (precipitation and temperature), and the determinants of adaptation to long-term global warming. Policymakers are generally more interested in the development of adaptation measures following political rather than hydrologic boundaries. Consequently, vulnerability and adaptation measures were also developed at the province and state levels for these two countries. In parallel, stakeholder forums were held in Ethiopia and South Africa to discuss measures of vulnerability, adaptation options and constraints, and the role of information and various actors—that is, the State, private sector, and civil society—in shaping adaptation to climate change. Finally, the impact of climate change on crop production in the survey sites was simulated based on crop yield and production function models to assess the implications of climate change for local food security.

At the basin level, the impact of climate change on water availability, water demands, and irrigation was simulated to identify basin-level adaptation strategies. Moreover, alternative investment strategies at the basin level were identified for Ethiopia taking into account climate variability and change, and broader impacts on the economy. A different but similar approach was used to study the impact of climate change and adaptation strategies on river basin units in South Africa. To capture the interactions of climate change and adaptation at the national and regional (Sub-Saharan Africa) levels, a water and food projections model was updated to take into account the impacts of climate change in addition to other drivers of global change. Using the integrated analysis tool, the impact of global change on poverty and water and food security was assessed for case study countries and Sub-Saharan Africa. Alternative adaptation strategies developed at workshops were assessed using the modeling framework, taking into account the local-level constraints and basin-level challenges identified.

These sets of analyses were complemented with papers on the role of climate change mitigation for the region, the importance of taking risk into account in devising adaptation options, and the role of collective action and property rights in community adaptation. The outcomes of the analyses can be used to guide appropriate response options to reduce rural vulnerability to global change.

CONCLUSIONS

The development of adaptive capacity to reduce adverse impacts of global change in rural areas of developing countries requires analyses at various spatial scales and an understanding of the linkages across the various scales. At the farm level, households adjust to global change by changing farm practices or abandoning farming. These local actions, in turn, influence climate and global change. At the basin level, basin authorities influence both land and water allocation, and carry out purposeful adaptations to global change. Purposeful adaptation can be either tactical, in response to climate or other global changes, or strategic, in anticipation of future global change. At the national level, governments and authorities influence ecosystem services and human well-being. They also carry out purposeful adaptations, including changes in price, trade, and investment policies to anticipate or respond to global change. At the regional level, organizations and institutions have the potential to mitigate global change impacts through changes in trading regimes, and the development of regional transportation and communications infrastructure. Important global factors also affect water and food security at the local level, such as world food trade and competition for water generated by the world economy.

This series of briefs describes the results of research in all these areas. While these results reflect only one round of research using the "strategic cyclical scaling" framework, it is hoped that future research will lead to even greater integration of bottom-up and top-down approaches.

FOR FURTHER READING

Ringler, C., The impact of climate variability and climate change on water and food outcomes: A framework for analysis, in C. van Bers, D. Petry, and C. Pahl-Wostl, eds., Global assessments: Bridging scales and linking to policy, GWSP Issues in Global Water System Research No. 2, http://www.gwsp.org/downloads/gwsp_issues_no2.pdf, 2007.

Root, T. L., and S. H. Schneider, Ecology and climate: Research strategies and implications. *Science* 269 (5222: 334–341, 1995).

C. Ringler (c.ringler@cgiar.org) is a senior research fellow at the International Food Policy Research Institute, Washington, DC.

This series of IFPRI Research Briefs is based on research supported by the Federal Ministry for Economic Cooperation and Development, Germany, under the project "Food and Water Security under Global Change: Developing Adaptive Capacity with a Focus on Rural Africa," which forms part of the CGIAR Challenge Program on Water and Food. Through collaboration with the Center for Environmental Economics and Policy in Africa, the Ethiopian Development Research Institute, the Ethiopian Economics Association, and the University of Hamburg, the project aims to provide policymakers and stakeholders in Ethiopia and South Africa with tools to better understand and analyze the consequences of global change—in particular climate change—and to form policy decisions that facilitate adaptation in these countries and beyond.

Financial Contributors and Partners

IFPRI's research, capacity strengthening, and communications work is made possible by its financial contributors and partners. IFPRI receives its principal funding from governments, private foundations, and international and regional organizations, most of which are members of the Consultative Group on International Agricultural Research (CGIAR). IFPRI gratefully acknowledges the generous unrestricted funding from Australia, Canada, China, Finland, France, Germany, India, Ireland, Italy, Japan, Netherlands, Norway, South Africa, Sweden, Switzerland, United Kingdom, United States, and World Bank.

Printed on alternative-fiber paper manufactured from agriculturally sustainable resources that are processed chlorine-free (PCF).

INTERNATIONAL FOOD POLICY RESEARCH INSTITUTE
sustainable solutions for ending hunger and poverty
Supported by the CGIAR

Center for Environmental Economics and Policy in Africa

IFPRI Research Brief 15–2

HOW CAN AFRICAN AGRICULTURE ADAPT TO CLIMATE CHANGE? INSIGHTS FROM ETHIOPIA AND SOUTH AFRICA

Vulnerability and the Impact of Climate Change in South Africa's Limpopo River Basin

Sharon Shewmake

With likely long-term changes in rainfall patterns and shifting temperature zones, climate change is expected to increase the frequency of climate-related shocks, such as floods and droughts in Sub-Saharan Africa. For farm households, an increase in the frequency of climate-related income shocks could lead not only to lower expected income, but also to higher income variance, which in turn can cause them to pursue costly risk-coping strategies and to fall below poverty trap thresholds. For this reason, it is important to understand how a changing distribution of risk will lead to increased vulnerability, not just decreased expected income.

This brief is based on a paper that uses farm households' responses to climate-related shocks in South Africa's Limpopo River Basin to gauge how farmers are likely to respond to future climate change shocks. The increased frequency and severity of droughts associated with global warming may induce farmers to change their behavior to reduce their risk of negative impacts from future climate-related shocks. While it is difficult to predict farmers' behavior and how it will affect welfare, coping strategies used in response to shocks today can be used to predict whether and how rural South Africans will cope in the future. The study examines household responses to droughts in 2005 and household vulnerability to future climate change by assessing their probability of falling below an income threshold as a result of future climate-related shocks.

Assessing household vulnerability is important because efficient social policy needs to go beyond poverty alleviation in the present to prevent poverty in the future. A poverty reduction strategy that does not distinguish between transient and structural poverty leads to inefficient use of resources, in that it focuses on those who are temporarily poor but may be able to escape poverty on their own, at the expense of those who have a high probability of becoming chronically poor.

IMPACT OF DROUGHTS ON INCOME

Using cross-sectional data from a 2005 survey of nearly 800 farmers in 20 districts in South Africa's Limpopo River Basin that was augmented with rainfall data, the study finds that droughts were the most prevalent climate-related shock reported: 41 instances of droughts were reported in 2005. No apparent relationship was found, however, between those who would have been expected to report droughts based on rainfall data and those who did report droughts. This may be because the rainfall variable examined is an average over the entire growing season and, as a result, may not capture a run of rainless days or other important microclimate dynamics.

In order to cope with shocks, the survey asked whether or not households sold livestock, borrowed from relatives or the bank, accepted aid, migrated to another area, sought off-farm employment, or ate less. The majority of the households, however, said they did nothing in response to droughts, and a corresponding regression analysis finds the impact of droughts on income to be insignificant.

This result is puzzling because, at the very least, households must have used their available assets or reduced their consumption in response to the decrease in income caused by climate-related shocks. There are two explanations for this. First, there could be problems with the data. Income may not have been accurately measured or the reports of no response could indicate that households reported many types of droughts, not just those that were severe and required coping strategies. Treating inconsequential droughts in the same way as severe droughts would dilute the signal of severe droughts on income. The second possible explanation is that households had already adapted to living in a drought-prone environment. Droughts are common in South Africa and households may already use drought-resistant crop varieties or other coping mechanisms that lessen their impact. This does not mean the presence of droughts does not affect household wealth (that is, their asset position). The coping strategies used by farmers in the presence of drought could be very costly and a reduction in income variance could increase overall income by allowing farmers to spend less time on low-yield, low-risk activities.

ESTIMATION OF VULNERABILITY

Vulnerability is defined as the probability that a household will fall below some income threshold at some future point in time. Using median household income (16,000 Rand, equivalent to US$2,508 in 2005) and bottom quartile income (7,800 Rand) as the two thresholds, the study assesses which households are more vulnerable to future climatic shocks. The results are plotted in four quadrants in Figures 1 and 2. Households in the upper-left quadrant are currently below the income threshold and will likely continue to be below the threshold in the future. Households in the bottom-left quadrant are currently below the threshold but have characteristics suggesting they have less than a 50 percent chance of being below the threshold in the future. Those in the upper-right corner are above the income threshold at present but are likely to fall below it in the future, whereas those in the bottom-right quadrant are above the income threshold and are likely to remain so in the future.

The results suggest that most households are not vulnerable to falling below the 7,800 rand threshold (Figure 1). However, when the poverty line is set at 16,000 Rand (the second income quartile), vulnerability increases significantly (Figure 2).

Further, the study finds that households that do not own farm animals are more vulnerable to falling below the poverty threshold than other households. Similarly, households with more than 10 members, and those that rely heavily on rainfed agriculture are more vulnerable to falling below the poverty threshold. Households that do not have access to credit or that have no knowledge of credit are vulnerable to falling below the poverty thresholds, but they are less vulnerable than households that succeeded in obtaining a loan.

The study also finds that in South Africa's Limpopo River Basin, residents of Gauteng province are most vulnerable to falling below the lower income threshold (7,800 Rand), whereas residents of Limpopo are the least vulnerable to falling below this threshold. Members of the siSwati and Setswana ethnic groups are the most vulnerable to falling below either poverty threshold.

Figure 1 Vulnerability to climate-related shocks based on the probability of household income falling below 7,800 rand

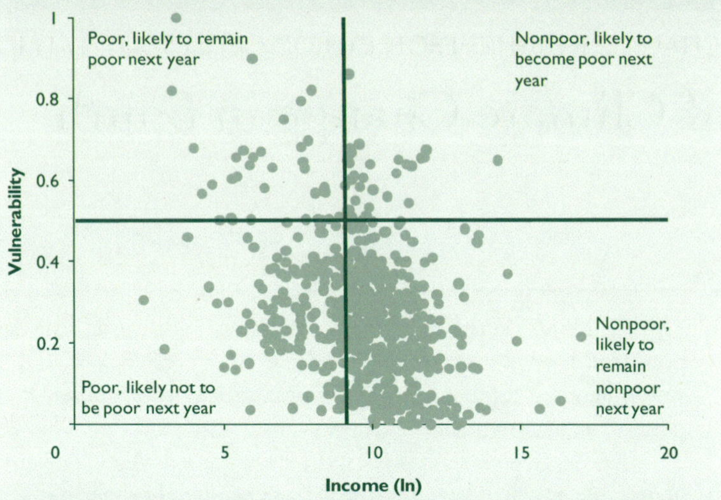

Figure 2 Vulnerability to climate-related shocks based on the probability of household income falling below 16,000 rand

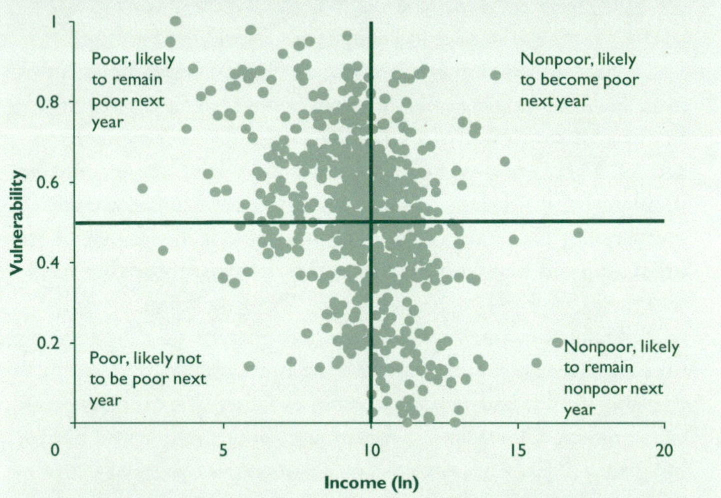

SOURCE: Shewmake, S., *Vulnerability and the Impact of Climate Change in South Africa's Limpopo River Basin*, IFPRI Discussion Paper No. 804 (Washington, DC: IFPRI, 2008).

IMPLICATIONS

While the impact of droughts on income was found to be statistically insignificant, and the majority of households said they "did nothing" in response to droughts, this does not necessarily mean that farming households in South Africa's Limpopo River Basin are prepared for future changes in climate patterns. Climate change is expected to bring a number of unexpected climate-related shocks requiring new adaptive behavior to mitigate their impacts. As a result, adaptation could be very costly, especially for those least able to bear that cost.

The vulnerability break-downs can help policymakers identify households that are not currently poor but are at risk of becoming poor in the future. Given that climate change will involve a redistribution and intensification of risk, attention to vulnerability is important. Given that predicting the actual effects of climate change is fraught with difficulties due to high levels of uncertainty, identifying households that are vulnerable to climate stresses will help to reduce future poverty.

FOR FURTHER READING

Shewmake, S., *Vulnerability and the Impact of Climate Change in South Africa's Limpopo River Basin*, IFPRI Discussion Paper No. 804, http://www.ifpri.org/pubs/dp/ifpridp00804.asp (Washington, DC: International Food Policy Research Institute, 2008).

S. Shewmake (shewmake@primal.ucdavis.edu) is a Ph.D. Candidate in the Department of Agricultural and Resource Economics at the University of California, Davis.

This series of IFPRI Research Briefs is based on research supported by the Federal Ministry for Economic Cooperation and Development, Germany, under the project "Food and Water Security under Global Change: Developing Adaptive Capacity with a Focus on Rural Africa," which forms part of the CGIAR Challenge Program on Water and Food. Through collaboration with the Center for Environmental Economics and Policy in Africa, the Ethiopian Development Research Institute, the Ethiopian Economics Association, and the University of Hamburg, the project aims to provide policymakers and stakeholders in Ethiopia and South Africa with tools to better understand and analyze the consequences of global change—in particular climate change—and to form policy decisions that facilitate adaptation in these countries and beyond.

Financial Contributors and Partners

IFPRI's research, capacity strengthening, and communications work is made possible by its financial contributors and partners. IFPRI receives its principal funding from governments, private foundations, and international and regional organizations, most of which are members of the Consultative Group on International Agricultural Research (CGIAR). IFPRI gratefully acknowledges the generous unrestricted funding from Australia, Canada, China, Finland, France, Germany, India, Ireland, Italy, Japan, Netherlands, Norway, South Africa, Sweden, Switzerland, United Kingdom, United States, and World Bank.

Printed on alternative-fiber paper manufactured from agriculturally sustainable resources that are processed chlorine-free (PCF).

INTERNATIONAL FOOD POLICY RESEARCH INSTITUTE
sustainable solutions for ending hunger and poverty
IFPRI®
Supported by the CGIAR

CEEPA
Center for Environmental Economics and Policy in Africa

IFPRI Research Brief 15–3

HOW CAN AFRICAN AGRICULTURE ADAPT TO CLIMATE CHANGE? INSIGHTS FROM ETHIOPIA AND SOUTH AFRICA

Mapping the South African Farming Sector's Vulnerability to Climate Change and Variability

A Subnational Assessment

Glwadys Aymone Gbetibouo and Claudia Ringler

In southern Africa, by the middle of the 21st century climate change is expected to cause temperature increases of 1–3°C, broad summer rainfall reductions of 5–10 percent, and an increase in the incidence of both droughts and floods. Consequently, climate change has significant potential to negatively affect crop production in South Africa, and in turn the well-being of the country's farmers.

This brief is based on a study that examines the level of vulnerability to climate change in South Africa's farming sector by developing a nationwide provincial-level vulnerability profile. Particular attention is paid to the underlying socioeconomic and institutional factors that determine how farmers respond to and cope with climate hazards.

CONCEPTUAL FRAMEWORK

In this study, vulnerability to climate change is conceptualized as a function of three factors: exposure, sensitivity, and adaptive capacity. Exposure can be interpreted as the direct danger (the stressor) together with the nature and extent of changes in a region's climate variables (temperature, precipitation, and extreme weather events). Sensitivity describes the human–environmental conditions that exacerbate or ameliorate the hazard, or trigger an impact. Exposure and sensitivity are intrinsically linked and mutually influence potential impacts. Adaptive capacity represents the potential to implement adaptation measures in efforts to avert potential impacts (Figure 1). Several indicators representing these three components were selected to facilitate the study's examination of vulnerability in South Africa. The selected indicators—drawn from an extensive review of the literature—represent both the biophysical conditions of the farming regions and the socioeconomic conditions of the farmers.

RESULTS OF VULNERABILITY ASSESSMENT

Analysis of vulnerability indicators shows that provinces in South Africa demonstrate vast diversity in environmental and socioeconomic conditions. The coastal provinces of the Eastern Cape, KwaZulu-Natal, and the Western Cape exhibit the highest frequency of extreme events (droughts and floods) over the past century, whereas the desert region of the Northern Cape and the steppe arid regions of the North West and Free State provinces exhibit the lowest frequency. The highest incremental temperature increase by 2050 is found in the desert region of the Northern Cape and the steppe arid regions of Free State and Mpumalanga, and rainfall changes are predicted to be greatest in the Gauteng and North West provinces.

The most sensitive provinces—mainly due to their high proportion of smallholder subsistence farmers—are the Eastern Cape, KwaZulu-Natal, and Limpopo. Smallholder farmers constitute 70 percent of the farming population in the Eastern Cape, KwaZulu-Natal, Mpumalanga, and the North West province, and inappropriate land uses in these regions have severely degraded land and reduced production capacity. The Eastern Cape and Limpopo provinces have the highest shares of agricultural GDP, the lowest average farm-asset values, the lowest literacy rates, and the highest unemployment rates.

The least-sensitive provinces are the Western Cape, Gauteng, and Free State. A common feature of these regions is that they have a low percentage of subsistence farmers and the least-populated rural areas. Gauteng and the Western Cape have greater infrastructure development, high levels of literacy, and lower unemployment rates. The Western Cape is the least sensitive province, largely due to a high degree of crop diversification, low levels of land degradation, and high reliance on irrigation.

Combining the indicators for sensitivity and exposure, KwaZulu-Natal, Limpopo, and the Eastern Cape are predicted to suffer the largest impacts of climate change and variability. With the exception of Limpopo, these provinces have both the largest exposure

Figure 1 Vulnerability framework

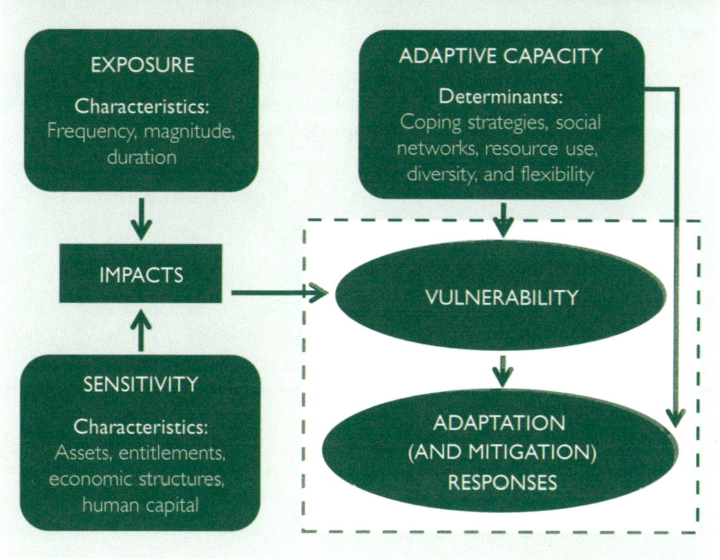

and the highest sensitivity. The Mpumalanga and North West provinces fall within the mid range of vulnerability based on these two indicators, whereas the Northern Cape, Western Cape, Free State, and Gauteng have the lowest risk of negative impacts because of the prevalence of commercial farming and the lack of land degradation.

Indicators of adaptive capacity differ considerably across the nine provinces. Capacity is greatest in the Western Cape due to the combined effects of well-developed infrastructure, high literacy rates and income levels, low unemployment rates and HIV prevalence, and relatively high capital wealth. Gauteng and the Northern Cape fall within the mid-range for this indicator, whereas adaptive capacity is low in KwaZulu-Natal, the Eastern Cape, Free State, Limpopo, and North West due to high dependence on agriculture, high unemployment rates and HIV prevalence, and low levels infrastructure development.

Assessing results based on all three components of vulnerability, Limpopo, the Eastern Cape and KwaZulu-Natal are the most vulnerable provinces; Mpumalanga, the North West, Gauteng, and the Northern Cape fall within the mid range of vulnerability; and the Western Cape has the lowest level of vulnerability. The vulnerability of Free State is considered indeterminate because it exhibits both low exposure and low adaptive capacity.

Figure 2 Vulnerability ranking in South Africa by province

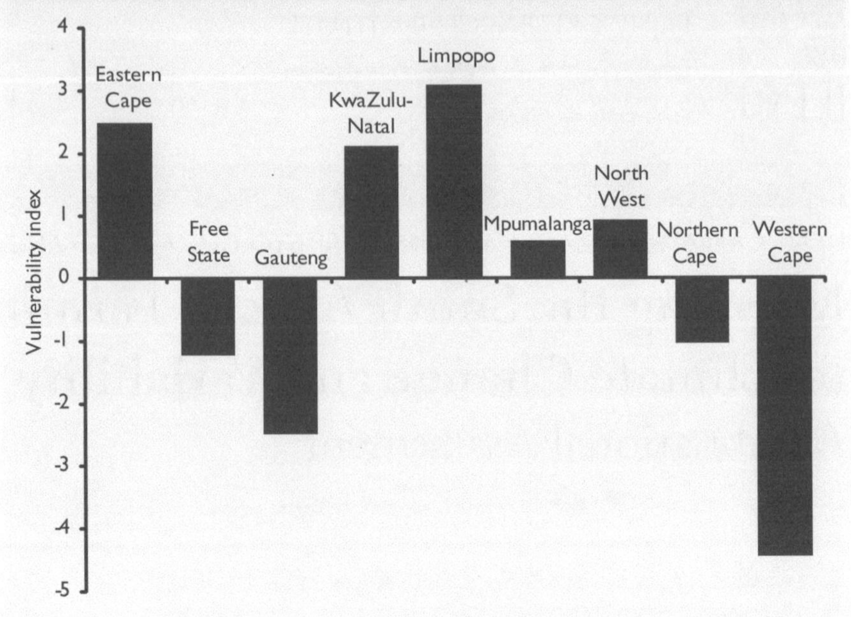

Figure 2 presents the results of a quantitative vulnerability index based on 19 indicators. As expected, the Western Cape and Gauteng have low vulnerability scores; the Free State, Northern Cape, Mpumalanga, and North West provinces fall within the mid range of vulnerability; and the most vulnerable provinces are the Eastern Cape, KwaZulu-Natal and Limpopo.

POLICY IMPLICATIONS

In examining vulnerability at the province level, caution must be taken given enormous heterogeneity in household-level resource access, poverty levels, and adaptive capacity. Ideally, future household-level research will facilitate improved targeting of policies to reduce climate change vulnerability. That said, the results of this study show that the provinces deemed most vulnerable to the effects of climate change and variability do not always equate with the most vulnerable populations. Rather, results suggest that the overall vulnerability of the South African farming sector is characterized by a combination of medium-level exposure risk coupled with medium to high levels of social vulnerability.

In light of large spatial differences in vulnerability, policymakers should tailor policies to local conditions. In highly vulnerable regions, such as Limpopo, KwaZulu-Natal, and the Eastern Cape, policymakers should enact measures (1) to support the effective management of environmental resources (for example, soil, vegetation, and water resources); (2) to promote increased market participation, especially within the large subsistence-farming sector; (3) to stimulate both agricultural intensification and livelihood diversification away from risky agriculture; and (4) to enact social programs and spending on health, education, and welfare to help maintain and augment both physical and intangible human capital. Policymakers should also invest in rural infrastructure development. In areas of high exposure, such as the coastal zones, priority should be given to the development of accurate early warning systems, as well as appropriate relief programs and agricultural insurance.

FOR FURTHER READING

Gbetibouo, G.A. and C. Ringler. *Mapping South African farming sector vulnerability to climate change and variability: A subnational assessment.* IFPRI Discussion Paper No. 885 (Washington, DC: International Food Policy Research Institute, 2009).

Glwadys Aymone Gbetibouo (gbetibouo@yahoo.fr) is a PhD student at the Center for Environmental Economics and Policy Analysis (CEEPA) of the University of Pretoria, South Africa. **Claudia Ringler** (c.ringler@cgiar.org) is a senior research fellow in the Environment and Production Technology Division (EPTD) of the International Food Policy Research Institute (IFPRI), Washington, DC.

This series of IFPRI Research Briefs is based on research supported by the Federal Ministry for Economic Cooperation and Development, Germany, under the project "Food and Water Security under Global Change: Developing Adaptive Capacity with a Focus on Rural Africa," which forms part of the CGIAR Challenge Program on Water and Food. Through collaboration with the Center for Environmental Economics and Policy in Africa, the Ethiopian Development Research Institute, the Ethiopian Economics Association, and the University of Hamburg, the project aims to provide policymakers and stakeholders in Ethiopia and South Africa with tools to better understand and analyze the consequences of global change—in particular climate change—and to form policy decisions that facilitate adaptation in these countries and beyond.

Financial Contributors and Partners

IFPRI's research, capacity strengthening, and communications work is made possible by its financial contributors and partners. IFPRI receives its principal funding from governments, private foundations, and international and regional organizations, most of which are members of the Consultative Group on International Agricultural Research (CGIAR). IFPRI gratefully acknowledges the generous unrestricted funding from Australia, Canada, China, Finland, France, Germany, India, Ireland, Italy, Japan, Netherlands, Norway, South Africa, Sweden, Switzerland, United Kingdom, United States, and World Bank.

Printed on alternative-fiber paper manufactured from agriculturally sustainable resources that are processed chlorine-free (PCF).

INTERNATIONAL FOOD POLICY RESEARCH INSTITUTE
sustainable solutions for ending hunger and poverty
IFPRI®
Supported by the CGIAR

EDRI
Ethiopia Development Research Institute

IFPRI Research Brief 15–4

Assessing Household Vulnerability to Climate Change

The Case of Farmers in the Nile Basin of Ethiopia

Temesgen T. Deressa, Rashid M. Hassan, and Claudia Ringler

Ethiopia remains one of the least-developed countries in the world: 50 percent of the population lives in abject poverty, and average life expectancy is only 43 years. Agriculture—the main sector of the Ethiopian economy—employs about 80 percent of the population and is dominated by small-scale, mixed crop and livestock production with very low productivity, which can be attributed to obsolete farming techniques; soil degradation caused by overgrazing and deforestation; poor complementary services, such as extension, credit, markets, and infrastructure; and frequent droughts and floods. Climate change is expected to increase the frequency, intensity, and spatial coverage of droughts and floods. Assessing Ethiopian farmers' vulnerability to climate change can help identify groups in particular need of support under a changing climate. This brief is based on a study that measures the vulnerability of farmers to climatic extremes, such as droughts, floods, and hailstorms.

THE IMPACT OF DROUGHT ON HOUSEHOLDS

The study found that of the 1,000 households surveyed across 5 regional states and 20 districts of Ethiopia, 31 percent reported droughts, 12 percent reported floods, and 18 percent reported hailstorms over the previous five years. These shocks resulted in a variety of reported losses, primarily in the form of crop yield declines and asset/income losses. The majority of farmers did nothing in response to these shocks, mainly due to abject poverty. Among those farmers who did respond, the most common coping strategy was to sell livestock, suggesting that—in addition to providing animal draft power and manure—livestock also serves as an asset and hence provides insurance against shocks. Other important coping strategies reported were borrowing from relatives, reducing food consumption, depending on food aid and food-for-work programs, and seeking off-farm employment.

ESTIMATION OF VULNERABILITY

Vulnerability is defined as the probability that climate shocks will shift household income below a given minimum level (such as a poverty line) or cause income levels to remain below the minimum level if the household is already poor. To analyze the sensitivity of vulnerability to a chosen threshold, the study examined different minimum income levels/poverty lines, such as the international poverty line of US$1.25 per day, the average income of the surveyed households, and arbitrary values above and below the average income of the surveyed households.

Results for income thresholds of US$0.3 and US$2 per day are provided in Figures 1 and 2, respectively. The x-axis shows current income, while the y-axis shows estimates of vulnerability. Each figure is divided into four quadrants. Farm households in the upper left quadrant fall below the relevant poverty line and are likely to remain there in the future; those in the bottom left quadrant fall below the poverty threshold today but have characteristics suggesting less than a 50 percent likelihood of their remaining there in the future; those in the upper right quadrant are currently above the specified income threshold but are likely to fall below it in the future; and those in the bottom left quadrant are currently above the income threshold and are likely to remain there in the future.

Unsurprisingly, farmers' vulnerability to climate shocks is highly sensitive to the choice of poverty threshold. The number of households that are considered poor today and likely to remain poor in the future increases as the income level used to determine the poverty threshold increases. When the poverty line is fixed at an income level of US$2 per day, 99 percent of farmers are considered vulnerable and, hence, fall into the upper right quadrant (Figure 1). Similarly, when the poverty line is reduced to US$1.50 and to US$1.25 per day, most households remain poor and vulnerable today, but the number of nonvulnerable households increases slightly. When the poverty threshold is set at a daily minimum income of US$0.30, only 12 percent of farmers are considered vulnerable to climate extremes (Figure 2). At this threshold, most households fall within the bottom left quadrant, indicating that they are not considered poor and are likely to remain nonpoor in the future.

Analysis of vulnerability across different agroecological zones indicates that farmers in the warm, semi-arid Kola zone are most vulnerable to extreme climatic events: of all the households surveyed in Kola, 99 percent are either already vulnerable or will become vulnerable in the future. Farmers in the cool, subhumid Weynadega zone and the cool, humid Dega zone are also highly vulnerable to future climate shocks: 94 and 89 percent, respectively, are considered vulnerable based on the US$1.25 per day poverty line. While these results are not representative at the regional level,

Figure 1 Vulnerability to climate shocks based on an income threshold of US$2 per day or 6,570 Ethiopian Birr per year

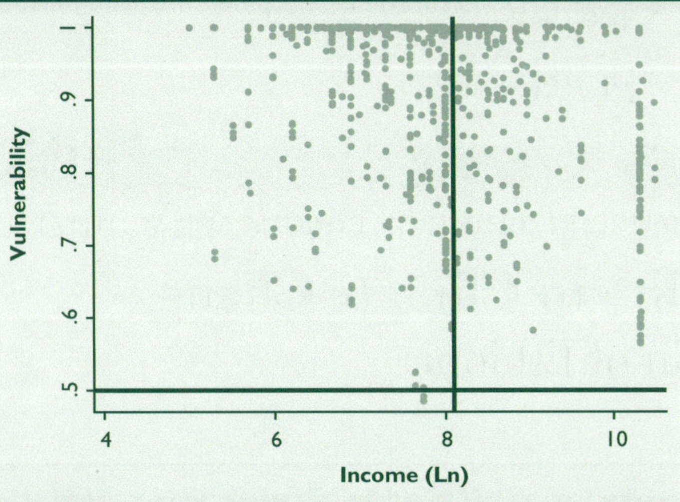

Figure 2 Vulnerability to climate shocks based on an income threshold of US$0.30 per day or 900 Ethiopian Birr per year

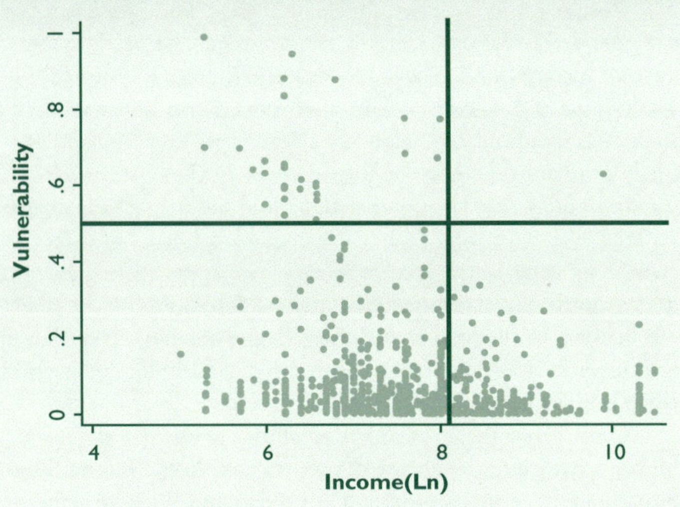

they do show that the districts surveyed in Beneshangul Gumuz and the Southern Nations, Nationalities and Peoples Region are most vulnerable under all of the poverty lines tested, whereas the districts surveyed in Oromia are the least vulnerable.

CONCLUSIONS AND POLICY IMPLICATIONS

The results of this study indicate an inverse relationship between the incomes of farm households and their level of vulnerability to climate extremes. Thus, policy interventions under climate change should focus on strengthening public and household-level risk management using both mitigation and adaptation strategies to reduce the negative impacts of climate change. Household-level mitigation strategies should include practices that encourage crop and livestock diversification, the use of drought-tolerant crop varieties and livestock species, mixed crop and livestock production, and membership in rotating credit groups. Policies that support household-level adaptation strategies should encourage income generation and asset holding, both of which provide households with a financial buffer against harsh climatic events.

Public risk mitigation might include strategies such as water harvesting, resource conservation and management, irrigation development, voluntary resettlement programs, targeted extension service packages (focusing on households or agroecological zones), targeted safety net programs, weather-indexed drought insurance, and the development of well-coordinated early warning systems. Important public adaptation strategies include the efficient administration of foreign emergency relief and effective food-for-work programs.

FOR FURTHER READING

Deressa, T. T., R. M. Hassan, and C. Ringler. *Assessing Household Vulnerability to Climate Change: The Case of Farmers in the Nile Basin of Ethiopia*. IFPRI Discussion Paper No. 935 (Washington, DC: International Food Policy Research Institute, 2009).

Temesgen T. Deressa (tderessa@yahoo.com) is a PhD student at the Centre for Environmental Economics and Policy for Africa (CEEPA), University of Pretoria. **Rashid M. Hassan** (rashid.hassan@up.ac.za) is the director of CEEPA. **Claudia Ringler** (c.ringler@cgiar.org) is a senior research fellow at the International Food Policy Research Institute, Washington, DC.

This series of IFPRI Research Briefs is based on research supported by the Federal Ministry for Economic Cooperation and Development, Germany, under the project "Food and Water Security under Global Change: Developing Adaptive Capacity with a Focus on Rural Africa," which forms part of the CGIAR Challenge Program on Water and Food. Through collaboration with the Center for Environmental Economics and Policy in Africa, the Ethiopian Development Research Institute, the Ethiopian Economics Association, and the University of Hamburg, the project aims to provide policymakers and stakeholders in Ethiopia and South Africa with tools to better understand and analyze the consequences of global change—in particular climate change—and to form policy decisions that facilitate adaptation in these countries and beyond.

Financial Contributors and Partners

IFPRI's research, capacity strengthening, and communications work is made possible by its financial contributors and partners. IFPRI receives its principal funding from governments, private foundations, and international and regional organizations, most of which are members of the Consultative Group on International Agricultural Research (CGIAR). IFPRI gratefully acknowledges the generous unrestricted funding from Australia, Canada, China, Denmark, Finland, France, Germany, India, Ireland, Italy, Japan, the Netherlands, Norway, the Philippines, South Africa, Sweden, Switzerland, the United Kingdom, the United States, and the World Bank.

Printed on alternative-fiber paper manufactured from agriculturally sustainable resources that are processed chlorine-free (PCF).

INTERNATIONAL FOOD POLICY RESEARCH INSTITUTE
sustainable solutions for ending hunger and poverty
Supported by the CGIAR

EDRI
Ethiopia Development Research Institute

IFPRI Research Brief 15–5

HOW CAN AFRICAN AGRICULTURE ADAPT TO CLIMATE CHANGE? INSIGHTS FROM ETHIOPIA AND SOUTH AFRICA

Measuring Ethiopian Farmers' Vulnerability to Climate Change Across Regional States

Temesgen T. Deressa, Rashid M. Hassan, and Claudia Ringler

Ethiopia's agricultural sector, which is dominated by small-scale, mixed crop, and livestock farming, is the mainstay of the country's economy. It constitutes more than half the nation's gross domestic product (GDP), generates more than 85 percent of the foreign exchange earnings, and employs about 80 percent of the population. Ethiopia's dependence on agriculture makes the country particularly vulnerable to the adverse impacts of climate change on crop and livestock production.

This brief is based on a paper that analyzes the vulnerability of Ethiopian farmers to climate change by creating a vulnerability index and comparing vulnerability indicators across regions. A regional vulnerability index can assist in identifying the areas of Ethiopia that are most vulnerable to climate change and guide policymakers in determining where investments in adaptation may be most effective in reducing the future adverse effects of climate change.

VULNERABILITY

According to the Intergovernmental Panel on Climate Change (IPCC), a region's vulnerability to climate change depends on its adaptive capacity, sensitivity, and exposure to changing climatic patterns. Adaptive capacity describes the ability of a system to adjust to actual or expected climate impacts or to cope with the consequences of climate change. Sensitivity is the degree to which a system is affected—whether positively or negatively—by extreme weather conditions and associated climatic variations. Exposure refers to the degree to which a system is exposed to climate change and the nature of the climate stimulus.

The indicators chosen to reflect adaptive capacity include household wealth, access to and use of technology, availability of infrastructure and institutions, potential for irrigation, and literacy rates. Wealth enables communities to absorb and recover from losses more quickly. The livestock owned (number of heads), key assets owned (radios, refrigerators, and so on), and quality of residential homes are commonly used as indicators of wealth in rural African communities. Proximity to agricultural input supplies is identified as an indicator of technology. For instance, drought-tolerant or early maturing varieties of crops generally require access to complementary inputs, such as fertilizers and pesticides.

Well-developed institutions and infrastructure also play important roles in adapting to climate change by facilitating access to resources. For instance, all-weather roads facilitate the distribution of necessary inputs to farmers and increase access to markets. Health services enable the provision of preventive treatments for diseases associated with climatic change, such as malaria, and the availability of microfinance supports the adoption of technology packages.

Irrigation potential and literacy rates are important factors contributing to adaptive capacity. Locations with more potentially irrigable land can adapt to climate change through improved water control. Regions with a higher literacy rate—a proxy for the level of education—are considered to have greater adaptive capacity.

Generally, increased frequency of droughts and floods negatively affects agricultural production, demonstrating agriculture's sensitivity to climate change. Finally, a region's exposure to climate change is represented by the predicted change in temperature and rainfall by 2050.

INDICATORS OF VULNERABILITY ACROSS ETHIOPIA'S ADMINISTRATIVE REGIONS

The study examined indicators of vulnerability across 7 of the 11 regions of Ethiopia: Afar; Amhara; Beneshangul Gumuz; Oromia; the Southern Nations, Nationalities, and People's Region (SNNPR); Somali; and Tigray. Data were not available for the Gambella region; and the urban centers of Addis Ababa, Dire Dawa, and Harari were excluded.

Regarding the indicators of wealth across the study regions, farmers living in the Amhara and Oromia regions have better quality housing compared with other regions. However, the share of people owning a radio is highest in Afar and lowest in Amhara. Livestock ownership is highest in Somali, given that most of the region's farmers are nomads whose livelihoods depend on livestock. Overall, few farmers in Ethiopia have access to nonagricultural income, gifts, and remittances.

Among the regions studied, farmers in SNNPR have the greatest access to technology, pesticides, fertilizers, and improved seeds, while farmers in the Somali and Afar regions have the lowest access.

The Afar region has the highest proportion of all-weather roads and health services, whereas Somali has the lowest proportion of health services, and Amhara has the lowest proportion of all-weather roads. Market access is highest in SNNPR and lowest in Somali and Amhara. Primary and secondary schools are equally distributed across the regions except for Somali, which has proportionally fewer schools. Telephone service is most available in rural Afar and least available in Beneshangul Gumuz. Tigray has the highest proportion of microfinance and veterinary services, whereas Somali has the lowest proportion of both these services. Irrigation potential and literacy rates are highest in SNNPR and Tigray and lowest in Afar and Somali.

Over the past century, Amhara has suffered most from droughts and floods, with Oromia and Somali following closely behind, whereas Beneshangul Gumuz and Afar have experienced the lowest number of droughts and floods. The predicted increase in temperature by 2050 is greatest for Afar and Tigray and lowest for SNPPR, and the change in precipitation is highest for Somali and lowest for SNPPR.

Figure 1 Vulnerability indicators of seven regional states in Ethiopia

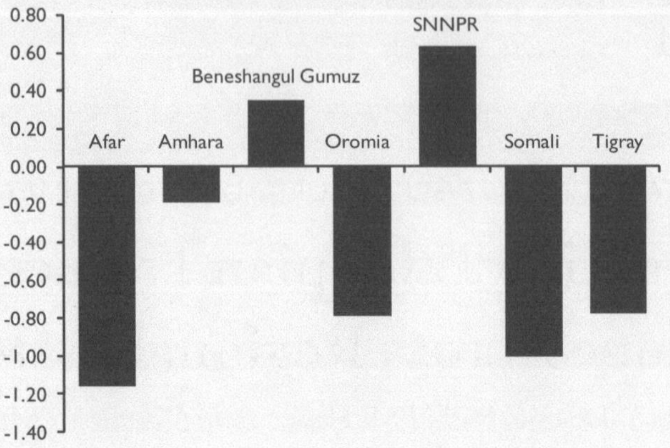

SOURCE: Deressa, T. T., R. M. Hassan, and C. Ringler, *Measuring Vulnerability of Ethiopian Farmers to Climate Change across Regional States*, IFPRI Discussion Paper (Washington, DC: IFPRI, 2008 forthcoming).
NOTE: Positive values indicate low vulnerability. Indexes are constructed using the principal component analysis method.

VULNERABILITY TO CLIMATE CHANGE BY REGION

The study calculated vulnerability to climate change by region as the net effect of sensitivity, exposure, and adaptive capacity. Results indicate that Afar, Somali, Oromia, and Tigray are relatively more vulnerable to climate change than the other regions. The vulnerability of Afar and Somali is attributed to their low level of rural service provision and infrastructure development. Tigray and Oromia's vulnerability to climate change can be attributed to the regions' higher frequencies of droughts and floods, lower access to technology, fewer institutions, and lack of infrastructure. SNNPR's lower vulnerability is associated with the region's relatively greater access to technology and markets, larger irrigation potential, and higher literacy rate.

POLICY IMPLICATIONS

While the results of the study identify the regions of Ethiopia most vulnerable to climate change, each region covers a large area with diverse socioeconomic and environmental characteristics. Future research should focus on uncovering localized vulnerability. Nonetheless, the findings from the regional study can help identify policies that may help stabilize national and regional food production in the face of the anticipated adverse effects of climate change.

In general, Ethiopia's vulnerability to climate change is highly correlated with poverty. Integrated rural development schemes aimed at alleviating poverty would increase the country's capacity to adapt to climate change. Policymakers would do well to prioritize poverty alleviation in the least-developed regions of the country, Afar and Somali, and the relatively more populated regions of Oromia and Tigray.

Likewise, investing in irrigation in areas with high potential, such as in SNNPR, could promote adaptation to climate change. Strengthening the ongoing micro-level adaptation programs implemented by governmental and nongovernmental organizations, such as water harvesting and other natural resource conservation programs, can also boost the adaptive capacities of farmers.

FOR FURTHER READING

Deressa, T. T., R. M. Hassan, and C. Ringler, *Measuring Ethiopian Farmers' Vulnerability to Climate Change Across Regional States*, IFPRI Discussion Paper No. 806, http://www.ifpri.org/pubs/dp/ifpridp00806.asp (Washington, DC: IFPRI, 2008).

T. T. Deressa (tderessa@yahoo.com) is a PhD student at the Centre for Environmental Economics and Policy for Africa (CEEPA), University of Pretoria. **R. M. Hassan** (rashid.hassan@up.ac.za) is director of CEEPA. **C. Ringler** (c.ringler@cgiar.org) is a senior research fellow at the International Food Policy Research Institute, Washington, DC.

This series of IFPRI Research Briefs is based on research supported by the Federal Ministry for Economic Cooperation and Development, Germany, under the project "Food and Water Security under Global Change: Developing Adaptive Capacity with a Focus on Rural Africa," which forms part of the CGIAR Challenge Program on Water and Food. Through collaboration with the Center for Environmental Economics and Policy in Africa, the Ethiopian Development Research Institute, the Ethiopian Economics Association, and the University of Hamburg, the project aims to provide policymakers and stakeholders in Ethiopia and South Africa with tools to better understand and analyze the consequences of global change—in particular climate change—and to form policy decisions that facilitate adaptation in these countries and beyond.

Financial Contributors and Partners

IFPRI's research, capacity strengthening, and communications work is made possible by its financial contributors and partners. IFPRI receives its principal funding from governments, private foundations, and international and regional organizations, most of which are members of the Consultative Group on International Agricultural Research (CGIAR). IFPRI gratefully acknowledges the generous unrestricted funding from Australia, Canada, China, Finland, France, Germany, India, Ireland, Italy, Japan, Netherlands, Norway, South Africa, Sweden, Switzerland, United Kingdom, United States, and World Bank.

Printed on alternative-fiber paper manufactured from agriculturally sustainable resources that are processed chlorine-free (PCF).

INTERNATIONAL FOOD POLICY RESEARCH INSTITUTE
sustainable solutions for ending hunger and poverty
Supported by the CGIAR

ETHIOPIAN ECONOMIC ASSOCIATION

IFPRI Research Brief 15–6

HOW CAN AFRICAN AGRICULTURE ADAPT TO CLIMATE CHANGE? INSIGHTS FROM ETHIOPIA AND SOUTH AFRICA

Perceptions of Stakeholders on Climate Change and Adaptation Strategies in Ethiopia

Assefa Admassie, Berhanu Adenew, and Abebe Tadege

The potential adverse effects of climate change on Ethiopia's agricultural sector are a major concern, particularly given the country's dependence on agricultural production. Securing Ethiopia's economic and social well-being in the face of climate change requires that policymakers and stakeholders work together to integrate climate change adaptation into the country's development process. Three stakeholder discussion forums held in 2006 in Addis Ababa, Awassa, and Bahir Dar as part of the project, "Food and Water Security under Global Change: Developing Adaptive Capacity with a Focus on Rural Africa," were attended by representatives of the government, civil society, business sector, and local communities. The forums elicited information to enable policymakers to make more informed decisions related to climate change adaptation.

The forums complemented ongoing efforts to develop the National Adaptation Program of Action (NAPA), which is overseen by a steering committee representing government, nongovernment, academic, and research institutions. Under NAPA, multidisciplinary technical working groups have been formed to assess the country's vulnerability to the adverse consequences of climate change, gauge current adaptation efforts, and identify ways in which public agencies could assist in minimizing the adverse impacts of climate change. In addition, two national and eight regional workshops were conducted involving nearly 500 participants with various areas of expertise. Like the stakeholder forums, the workshops solicited information to create greater awareness of climate change, assess the extent of the area's vulnerability, and help identify adaptation options.

This brief is based on a paper that presents findings from the stakeholder discussion forums, as well as NAPA's technical working groups and workshops. These meetings explored stakeholders' perceptions of vulnerability to climate change and considered ways in which adaptation measures could be further integrated into Ethiopia's development process.

VULNERABILITY TO THE ADVERSE CONSEQUENCES OF CLIMATE CHANGE

According to the National Meteorological Agency, long-term climate change in Ethiopia is associated with changes in precipitation patterns, rainfall variability, and temperature, which could increase the country's frequency of both droughts and floods. The stakeholder forums and NAPA's technical working groups sought participants' perceptions of the impacts of changing climatic conditions. According to participants, although both developed and developing countries are affected by climate change, developing countries face greater challenges in overcoming its adverse consequences. Because Ethiopia is one of the least developed countries in the world, with a per capita income of less than US$130 in 2006, workshop participants agreed that the country faces considerable hurdles in coping with the adverse impacts of long-term climate change. Low economic development, inadequate infrastructure, and lack of institutional capacity all contribute to the country's vulnerability to the adverse impacts of climate change.

In addition, Ethiopia's economy is heavily dependent on agriculture and faces increasing population growth. Ethiopia's agricultural sector contributes 47 percent of the country's gross national product and more than 80 percent of its exports. It also employs about 85 percent of the country's population of more than 76 million people (noting that Ethiopia is the third-most populous country in Africa after Nigeria and Egypt). With a current growth rate of about 2.8 percent per year, Ethiopia's population is expected to reach 129 million by 2030. Workshop participants emphasized that Ethiopia's low level of economic development combined with its heavy dependence on agriculture and high population growth rate make the country particularly susceptible to the adverse effects of climate change. Negative climatic impacts on crop and livestock production could result in a nationwide food shortage and greatly hinder the economy. If appropriate steps are not taken, workshop participants felt that food insecurity, deepened poverty, and increased incidence of disease, such as malaria and yellow fever, would be likely consequences.

A better understanding of the local dimensions of vulnerability is therefore essential to develop appropriate adaptation measures that will mitigate these adverse consequences. Accordingly, the stakeholder forums solicited input on those who are thought to be most vulnerable, based on economic, social, and environmental factors.

The farming community was identified as the most vulnerable because of its dependence on agricultural production for its livelihood. Within the farming community, small-scale, rainfed subsistence farmers as well as pastoralists were identified as more vulnerable to changing climatic conditions than others. In addition, farm households without assets and financial resources were identified as especially vulnerable, as their limited resources restrict them from easily adapting to the changing climate.

Participants noted that throughout society those who are marginalized based on their sex, age, education, ethnicity, or economic status are more vulnerable to negative climate change impacts than others. Female-headed households in Ethiopia, for instance, were identified as particularly vulnerable to climate change given greater constraints to adaptation than male-headed households. Likewise, participants argued that those with low levels of education are more vulnerable to the adverse effects of climate change because they are less likely to be aware of long-term climate change and tend to be more averse to the risks associated with some adaption measures.

In addition, certain regions of the country were identified as being more susceptible to the negative effects of changing climatic conditions. In particular, participants identified those in the arid, semi-arid, and dry subhumid lowlands as being more vulnerable to the adverse effects of climate change than those living in the highlands. Lowland households with fewer and less diversified assets than highland households are less able to invest in adaptation measures and to meet their basic needs when faced with the adverse consequences of climate change. In addition, the lowlands have become increasingly depleted and, as a result, many pastoralists from the lowlands have migrated to highland areas, leading to social conflicts over land.

ADAPTATION MEASURES AND POLICY RECOMMENDATIONS

Workshop participants identified the adaptation measures communities adopt when confronted with climate-related shocks. These include diversification of livelihood sources, migration, participation in nonfarm activities, sale of assets, settlement and resettlement activities, and the adoption of improved water management systems. To further mitigate the adverse effects of climate change, participants considered ways in which adaptation measures could be better integrated into Ethiopia's development process and offered policy recommendations.

As part of the development process, participants recommended that social and physical infrastructure be improved and institutions dealing with climate-related issues including the meteorology agency be strengthened to increase the country's adaptive capacity. In addition, workshop participants suggested implementing improved water resource development, land management, food security, health, and education programs. There is also a need to expand nonagricultural employment opportunities and provide skills training, particularly in rural areas.

Another recommendation is greater investment in data collection and research on climate change and extreme weather conditions in Ethiopia. Environmental and drought monitoring systems should be introduced at national and regional levels to monitor climate changes. Such systems would provide early warnings on predicted weather extremes, enabling stakeholders to take corrective measures in advance to minimize potential damages. Additionally, in-depth studies on vulnerability and adaptation should continue.

Participants also recommended expanding awareness of global warming and its potential impacts by providing reliable and up-to-date information to the public. Information about the appropriate adaptive measures should be made available to the entire national community. As part of this effort, communication between policymakers, nongovernmental organizations, research institutions, and the media, among other actors, should be strengthened in order to ensure accurate information is available and widely disseminated.

Finally, workshop participants recommended that all stakeholders be included in the development process. When considering measures to reduce vulnerability to the adverse effects of climate change, government, civil society, the private sector, and local communities should all participate in the discussion. In particular, policymakers should be sure to draw on knowledge and experience from local communities. By overlooking local knowledge, policies can constrain rather than enhance the adaptive capacity of communities. Furthermore, knowledge of the environment is passed down through generations of experience of working on the land and, thus, local farm communities offer invaluable information regarding adaptation to changing climatic conditions that would not likely be acquired through other channels.

FOR FURTHER READING

Admassie, A., and B. Adenew, Stakeholders' Perceptions of Climate Change and Adaptation Strategies in Ethiopia, *EEA Research Report* (Addis Ababa: Ethiopian Economic Association, 2008) Available at: http://www.eeaecon.org/Research%20Materials.htm.

A. Admassie (aadmassie@yahoo.com) is director of the Ethiopian Economic Association (EEA)/Ethiopian Economic Policy Research Institute (EEPRI) in Ethiopia. **B. Adenew** (berhanuad@yahoo.com) is a senior researcher at EEA/EEPRI in Ethiopia. **A. Tadege** (a_tadege@yahoo.com) is head of the Research and Studies Department at the National Meteorological Agency (NMA) in Ethiopia.

This series of IFPRI Research Briefs is based on research supported by the Federal Ministry for Economic Cooperation and Development, Germany, under the project "Food and Water Security under Global Change: Developing Adaptive Capacity with a Focus on Rural Africa," which forms part of the CGIAR Challenge Program on Water and Food. Through collaboration with the Center for Environmental Economics and Policy in Africa, the Ethiopian Development Research Institute, the Ethiopian Economics Association, and the University of Hamburg, the project aims to provide policymakers and stakeholders in Ethiopia and South Africa with tools to better understand and analyze the consequences of global change—in particular climate change—and to form policy decisions that facilitate adaptation in these countries and beyond.

Financial Contributors and Partners

IFPRI's research, capacity strengthening, and communications work is made possible by its financial contributors and partners. IFPRI receives its principal funding from governments, private foundations, and international and regional organizations, most of which are members of the Consultative Group on International Agricultural Research (CGIAR). IFPRI gratefully acknowledges the generous unrestricted funding from Australia, Canada, China, Finland, France, Germany, India, Ireland, Italy, Japan, Netherlands, Norway, South Africa, Sweden, Switzerland, United Kingdom, United States, and World Bank.

Printed on alternative-fiber paper manufactured from agriculturally sustainable resources that are processed chlorine-free (PCF).

INTERNATIONAL FOOD POLICY RESEARCH INSTITUTE
sustainable solutions for ending hunger and poverty
IFPRI®
Supported by the CGIAR

CEEPA
Center for Environmental Economics and Policy in Africa

IFPRI Research Brief 15–7

HOW CAN AFRICAN AGRICULTURE ADAPT TO CLIMATE CHANGE? INSIGHTS FROM ETHIOPIA AND SOUTH AFRICA

Micro-Level Analysis of Farmers' Adaptation to Climate Change in Southern Africa

Charles Nhemachena and Rashid M. Hassan

Agricultural production remains the main source of livelihood for rural communities in Sub-Saharan Africa, providing employment to more than 60 percent of the population and contributing about 30 percent of gross domestic product. With likely long-term changes in rainfall patterns and shifting temperature zones, climate change is expected to significantly affect agricultural production, which could be detrimental to the region's food security and economic growth. An assessment of the factors influencing farm-level adaptation can facilitate the formation of policies and investment strategies that help moderate potential adverse consequences of long-term climate change. Because smallholder farmers tend to have a low capacity to adapt to changes in climatic conditions, policies that help these farmers adapt to global warming and associated climatic extremes are particularly important.

This brief is based on a study that assesses smallholder farmers' adaptation to climate change in southern Africa. The study identifies farmers' perceptions of climate change and the determinants of farm-level adaptation strategies, and recommends policies that could help stabilize national and regional food production given the anticipated adverse effects of climate change.

SURVEY OF FARMERS' PERCEPTIONS

Using cross-sectional survey data for South Africa, Zambia, and Zimbabwe, the study finds that most farmers detect a rise in temperature over the past 20 years, drier conditions, and pronounced changes in the timing of rains and frequency of droughts. In response to these perceived changes in climate, 67 percent of survey respondents are adopting some form of adaptation. Common adaptation measures include diversifying crops, planting different crops or crop varieties, replacing farm activities with nonfarm activities, changing planting and harvesting dates, increasing the use of irrigation, and increasing the use of water and soil conservation techniques. In assessing farmers' perceptions of barriers to using various adaptation measures, the authors find that lack of credit, lack of information on climate, and insufficient access to inputs are key obstacles to adaptation (Figure 1).

DETERMINANTS OF FARMERS' USE OF ADAPTATION STRATEGIES

The study uses an econometric model to identify the factors that affect farmers' use of adaptation strategies. Modeling results confirm that awareness of climate change is an important determinant of farm-level adaptation. Access to credit, markets, and free extension services also significantly increase the likelihood of farmers adopting adaptation measures. In addition, households with access to electricity and technology such as tractors, heavy machines, and animal power are more likely to adapt to changes in climatic conditions. With access to electricity and technology, farmers are able to vary their planting dates, switch to new crops, diversify their crop options, use more irrigation, and apply water conservation techniques. Farmers with access to technology are also more likely to diversify into nonfarming activities, although households with large investments in farm equipment and machinery may find such diversification to be costly.

Another important determinant of farm-level adaptation is land ownership. Farmers who own their land are more likely to invest in adaptation options, including crop and livestock management practices and water conservation. The type of farming system also determines farmers' use of adaptation strategies: those engaged in mixed crop and livestock farming, as well as those engaged in subsistence farming, are more likely to adapt to changes in climatic conditions than are farmers in specialized farming systems.

Finally, the study finds that female-headed households are more likely to take up adaptation options than male-headed households. In most rural smallholder farming communities in South Africa, women do much of the agricultural work and therefore tend to have more farming experience and information on various management practices. Farming experience increases the probability of uptake of all adaptation options.

POLICY IMPLICATIONS

Providing smallholder farmers with necessary resources increases farmers' productivity and helps them adapt to the adverse consequences of changing climatic conditions. For instance, policies that ensure access to affordable credit increase farmers' financial resources, allowing them to make better use of available information on climate change and to meet the costs associated with the various adaptation options. Likewise, policies that ensure farmer access to free extension services have the potential to significantly increase farmers' awareness of changing climatic conditions and their knowledge of appropriate adaptation measures.

Because property owners are more likely to invest in adaptation measures, governments should also ensure that tenure arrangements are safeguarded, even in the communal systems that characterize most of the region's smallholder farming systems. Similarly, policies targeting women's groups and associations in smallholder rural communities could further promote adaptation given that women do much of the agricultural work in many rural smallholder farming communities in the region.

Finally, governments need to support research and development in the agricultural sector, disseminate appropriate technologies, and ensure that cheap technologies are available to smallholder farmers. Examples of these policy measures include developing drought-resistant crop technologies, improving the forecasting and dissemination of climate information, and promoting farm-level adaptation measures such as the use of irrigation technologies. Government policies designed to promote adaptation at the farm level will lead to greater food and livelihood security in the face of climate change.

FOR FURTHER READING

Nhemachena, C., and R. Hassan, *Micro-Level Analysis of Farmers' Adaptation to Climate Change in Southern Africa*, IFPRI Discussion Paper No. 714 (Washington, DC: International Food Policy Research Institute, 2007).

Figure 1 Perceived barriers to adaptation

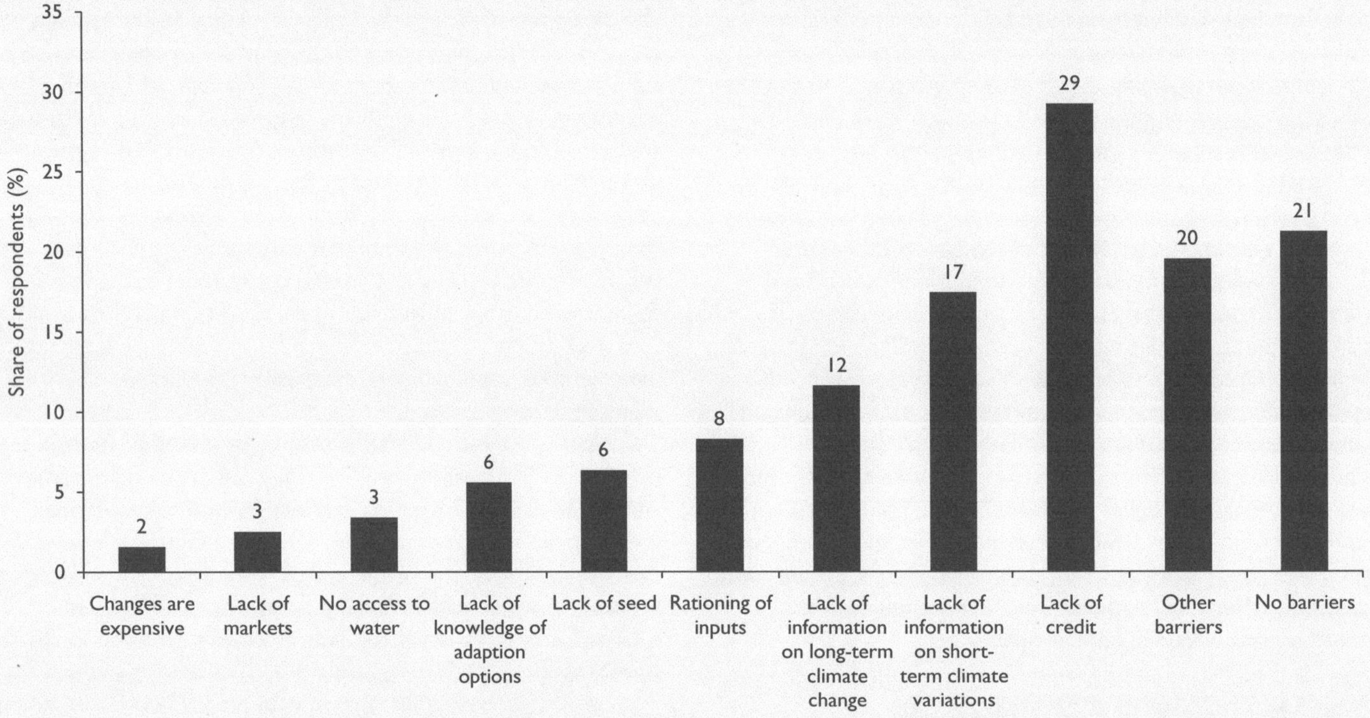

SOURCE: Nhemachena, C., and R. *Hassan, Micro-Level Analysis of Farmers' Adaptation to Climate Change in Southern Africa*, IFPRI Discussion Paper No. 714 (Washington, DC: International Food Policy Research Institute, 2007).

C. Nhemachena (nhemachenacharles@yahoo.co.uk) is a former Ph.D. student at the Centre for Environmental Economics and Policy in Africa (CEEPA). **R. M. Hassan** (rashid.hassan@up.ac.za) is the Director of CEEPA.

This series of IFPRI Research Briefs is based on research supported by the Federal Ministry for Economic Cooperation and Development, Germany, under the project "Food and Water Security under Global Change: Developing Adaptive Capacity with a Focus on Rural Africa," which forms part of the CGIAR Challenge Program on Water and Food. Through collaboration with the Center for Environmental Economics and Policy in Africa, the Ethiopian Development Research Institute, the Ethiopian Economics Association, and the University of Hamburg, the project aims to provide policymakers and stakeholders in Ethiopia and South Africa with tools to better understand and analyze the consequences of global change—in particular climate change—and to form policy decisions that facilitate adaptation in these countries and beyond.

Financial Contributors and Partners

IFPRI's research, capacity strengthening, and communications work is made possible by its financial contributors and partners. IFPRI receives its principal funding from governments, private foundations, and international and regional organizations, most of which are members of the Consultative Group on International Agricultural Research (CGIAR). IFPRI gratefully acknowledges the generous unrestricted funding from Australia, Canada, China, Finland, France, Germany, India, Ireland, Italy, Japan, Netherlands, Norway, South Africa, Sweden, Switzerland, United Kingdom, United States, and World Bank.

Printed on alternative-fiber paper manufactured from agriculturally sustainable resources that are processed chlorine-free (PCF).

INTERNATIONAL FOOD POLICY RESEARCH INSTITUTE
sustainable solutions for ending hunger and poverty
Supported by the CGIAR
IFPRI®

CEEPA
Center for Environmental Economics and Policy in Africa

IFPRI Research Brief 15–8

Understanding Farmers' Perceptions and Adaptations to Climate Change and Variability

The Case of the Limpopo Basin, South Africa

Glwadys Aymone Gbetibouo

Climate change is expected to have serious environmental, economic, and social impacts on South Africa. In particular, rural farmers, whose livelihoods depend on the use of natural resources, are likely to bear the brunt of adverse impacts. The extent to which these impacts are felt depends in large part on the extent of adaptation in response to climate change. Adaptation is widely recognized as a vital component of any policy response to climate change. Without adaptation, climate change would be detrimental to the agricultural sector, but with adaptation, vulnerability can be significantly reduced.

This brief is based on a study that examines farmers' perceptions of climate change and analyzes their adaptation responses to climate change and variability using household survey data from the Limpopo River Basin in South Africa.

FARMERS' PERCEPTIONS OF CLIMATE CHANGE

Farmers' ability to perceive climate change is a key precondition for their choice to adapt. The accuracy of farmers' perceptions of climate change was assessed by comparing their perceptions of long-term changes in temperature and precipitation with climate trends recorded at nearby meteorological stations. About 91 percent of the farmers surveyed perceived an increase in temperature over the past 20 years. This perception was confirmed by the statistical record for the Limpopo River Basin between 1960 and 2003, which showed the increase occurring mostly in the summer months (October to March). An analysis of climate data at the provincial level shows the same general trend of increasing temperature with some minor variations in terms of the severity of the increase and its timing (warming occurred mostly during the winter months in Limpopo, Gauteng, and Mpumalanga). Thus, farmers' perceptions are supported by the statistical record.

In terms of precipitation, 81 percent of respondents reported a decrease in rainfall over the past 20 years. Approximately 12 percent of farmers reported a change in the timing of rains, and many of these respondents observed a delayed and shorter rainfall season (summer).

The recorded rainfall data for 1960–2003 show that about 85 percent of rainfall occurs during summer months. However, despite a majority perception that rainfall had decreased—and with the exception of the winter season where the data do show a decreasing trend—the climate record shows no statistically significant trend over the past 40 years. Overall the climate record shows large variability in the amount of precipitation from year to year, and the same pattern was observed in each province. The high proportion of farmers noticing a decrease in precipitation could be due to the substantial decline in rainfall during 2001–03. Thus, farmer reports of a reduction in rainfall over the past 20 years may indicate that their perceptions are influenced by more recent climate trends.

A number of factors influence the likelihood that farmers will perceive climate change. Having fertile soil and access to water for irrigation decreases the likelihood that farmers will perceive climate changes, whereas education, experience, and access to extension services increase the likelihood that farmers will perceive climate changes. This suggests that perceptions are not based entirely on actual climate conditions and changes but are also influenced by other factors.

ADAPTATION TO CLIMATE CHANGE

Even though a large number of farmers interviewed noticed changes in climate, almost two-thirds chose not to undertake any remedial action. Among those farmers who did adapt, common responses included planting different crops, changing crop varieties, changing planting dates, increasing irrigation, diversifying crops, changing the amount of land grazed or under cultivation, and supplementing livestock feed. While adopting a new crop variety was the main strategy used to adapt to increasing temperature, building water-harvesting schemes was a popular strategy for coping with decreased precipitation.

Farmers cited a number of barriers to adaptation including poverty, lack of access to credit, and lack of savings. Insecure property rights and lack of markets were also cited as significant barriers to adaptation. A few farmers also reported lack of information and knowledge of appropriate adaptation measures as barriers to adaptation (Table 1).

Results from mathematical models indicate that experienced farmers are more likely to adapt to perceived climate change. In particular, the likelihood of diversifying portfolios (that is, adopting new crops or crop varieties, or using mixed farming systems), changing planting dates, and changing the amount of land under production increases with farm experience. This suggests that

Table 1 Barriers to adaptation in the Limpopo River Basin (% of the respondents)

	Lack of information about climate change	Lack of knowledge concerning appropriate adaptations	Poverty or lack of credit or savings	No access to water	Insecure property rights	Poor transport links or lack of market access	Others	No barriers
Total Basin	6	2	54	21	10	6	11	1
Limpopo	4	3	24	33	14	10	8	8
Northwest	10	0	55	3	3	1	9	22
Gauteng	0	0	32	12	0	4	20	10
Mpumalanga	9	2	48	9	6	1	13	23

farmers with greater management and technical skills are better able to cope with climate variability and change, and to spread risk by exploiting strategic complementarities among activities, such as crop–livestock integration.

Large households were more likely to adapt, indicating that additional household labor may facilitate the adoption of more labor-intensive adaptation options. Large-scale farmers were also more likely to adapt as a result of greater financial resources and access to capital, which enable them to invest in more costly technologies, such as irrigation. Likewise, wealthier households were more likely to change their planting dates in response to perceived climate change.

Access to rural services such as extension and credit also increases the likelihood of adaptation. Farmers with access to extension services—and who are therefore more likely to have knowledge of management practices to address climate changes—were more likely to diversify their portfolios in an effort to reduce risk. Access to credit also increases the likelihood that farmers will diversify their portfolios and buy feed supplements for their livestock. Given that lack of financial resources is one of the major constraints to adaptation, access to credit enables farmers with limited financial resources to purchase the necessary inputs and equipment associated with many adaptation options.

Having secure property rights also increases the probability of adaptation. With clear property rights farmers are able to make adaptation decisions involving their land, such as changing the amount of land under cultivation. Access to fertile soil also increases the likelihood that farmers will increase the amount of land under cultivation in response to climate change.

While off-farm employment could present a constraint to adaptation by reducing time spent managing farms, the results show that farmers who engage in off-farm activities are more likely to supplement livestock feed. This suggests that expanding smallholder farmers' access to off-farm income sources increases the probability that they will be able to afford adaptation measures.

The results show important regional variation. In the Limpopo province, which has a large rural population dependent on agriculture, farmers were more likely to adapt to climate changes compared with those in the others provinces. Similarly, households located in regions with high temperatures were more likely to adapt their farming practices, particularly by diversifying their portfolios, using irrigation, and changing planting dates.

CONCLUSIONS AND POLICY IMPLICATIONS

Given that few farmers adjusted their farming practices despite perceiving changes in climate, governments should facilitate adaptation by enabling farmers to overcome the barriers reported in this study. Specifically, policies should ensure that farmers have access to affordable credit, which would give them greater flexibility to modify their production strategies in response to climate change. Because access to water for irrigation increases farmers' resilience to climate variability, greater investments in smart irrigation are needed. Reforming pricing, clearly defining property rights, and strengthening farm-level managerial capacity should also be emphasized to promote efficient water use. More importantly, given that land reform has increased the number of less experienced and unskilled farmers, extension services need to be expanded with highly qualified personnel. Additional measures required are improving off-farm income-earning opportunities, and facilitating a smooth transition from subsistence to commercial farming.

FOR FURTHER READING

Gbetibouo, G. A., *Understanding Farmers' Perceptions and Adaptations to Climate Change and Variability: The Case of the Limpopo Basin, South Africa*, IFPRI Discussion Paper (Washington, DC: International Food Policy Research Institute, forthcoming 2009).

G. A. Gbetibouo (gbetibouo@yahoo.fr) is a Ph.D. student at the Center for Environmental Economics and Policy in Africa at the University of Pretoria, South Africa.

This series of IFPRI Research Briefs is based on research supported by the Federal Ministry for Economic Cooperation and Development, Germany, under the project "Food and Water Security under Global Change: Developing Adaptive Capacity with a Focus on Rural Africa," which forms part of the CGIAR Challenge Program on Water and Food. Through collaboration with the Center for Environmental Economics and Policy in Africa, the Ethiopian Development Research Institute, the Ethiopian Economics Association, and the University of Hamburg, the project aims to provide policymakers and stakeholders in Ethiopia and South Africa with tools to better understand and analyze the consequences of global change—in particular climate change—and to form policy decisions that facilitate adaptation in these countries and beyond.

Financial Contributors and Partners

IFPRI's research, capacity strengthening, and communications work is made possible by its financial contributors and partners. IFPRI receives its principal funding from governments, private foundations, and international and regional organizations, most of which are members of the Consultative Group on International Agricultural Research (CGIAR). IFPRI gratefully acknowledges the generous unrestricted funding from Australia, Canada, China, Finland, France, Germany, India, Ireland, Italy, Japan, Netherlands, Norway, South Africa, Sweden, Switzerland, United Kingdom, United States, and World Bank.

Printed on alternative-fiber paper manufactured from agriculturally sustainable resources that are processed chlorine-free (PCF).

INTERNATIONAL FOOD POLICY RESEARCH INSTITUTE
sustainable solutions for ending hunger and poverty
IFPRI®
Supported by the CGIAR

EDRI
Ethiopia Development Research Institute

IFPRI Research Brief 15–9

HOW CAN AFRICAN AGRICULTURE ADAPT TO CLIMATE CHANGE? INSIGHTS FROM ETHIOPIA AND SOUTH AFRICA

Analysis of the Determinants of Farmers' Choice of Adaptation Methods and Perceptions of Climate Change in the Nile Basin of Ethiopia

Temesgen T. Deressa, Rashid M. Hassan, Claudia Ringler, Tekie Alemu, and Mahmud Yesuf

Ethiopia's agricultural sector, which is dominated by small-scale, mixed-crop, and livestock farming, is the mainstay of the country's economy. It constitutes more than half of the country's gross domestic product, generates more than 85 percent of foreign exchange earnings, and employs about 80 percent of the population. Unfortunately, Ethiopia's dependence on agriculture makes the country particularly vulnerable to the adverse impacts of climate change on crop and livestock production. Thus, a deeper understanding of the complex interdependence between changing climatic conditions and Ethiopia's agricultural sector—together with adaptation options—is crucial.

Additional information about farmers' awareness of climate change and current adaptation approaches would assist policymakers in their efforts to decrease the country's vulnerability to the adverse impacts of climate change. This brief is based on a study that endeavors to guide policymakers on ways to promote greater adaptation by identifying the household characteristics that increase farmers' awareness of climate change and influence farmers' decision to adapt.

ADAPTATION METHODS

The study uses a cross-sectional household survey of farmers conducted during the 2004/05 production year in the Nile Basin of Ethiopia. The survey asked farmers how they respond to the negative impacts of climate change. Use of different crop varieties was the most common response, whereas use of irrigation was the least cited among the five adaptation methods identified. About 42 percent of those surveyed reported that they did not employ any adaptation methods (Figure 1).

Farmers reported that the primary reasons for choosing not to adapt were lack of information on climate change impacts and adaptation options, lack of financial resources, labor constraints, and land shortages. Adaptation to climate change is both costly and labor intensive. Lack of financial resources keeps farmers from acquiring the necessary technologies to allow them to adapt. For instance, while the Nile Basin in Ethiopia is rich in water resources, farmers generally cannot afford to invest in irrigation technology that would allow them to adapt to climate change or sustain their livelihoods during climatic extremes, such as drought. Additionally, farmers are often unable to mobilize sufficient family labor or afford hired labor to make the necessary changes. Moreover, high population pressures force farmers to intensively farm small plots of land, making it difficult to adopt adaptation practices, such as planting trees, which require more land.

Figure 1 Farmers' methods of adapting to climate change

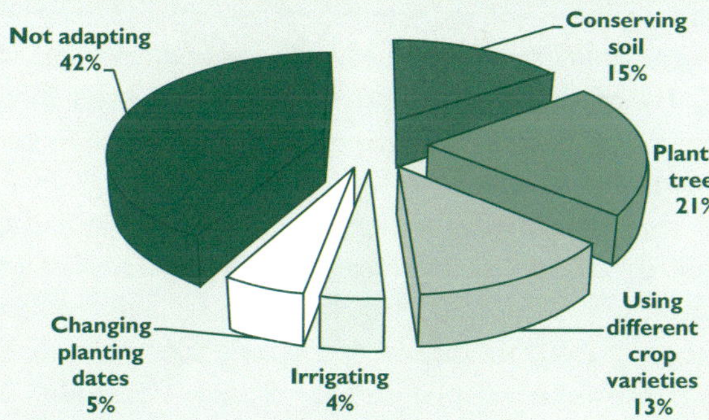

SOURCE: Deressa, T. T., R. M. Hassan, C. Ringler, T. Alemu, and M. Yesuf, *Analysis of the Determinants of Farmers' Choice of Adaptation Methods and Perceptions of Climate Change in the Nile Basin of Ethiopia*, IFPRI Discussion Paper (Washington, DC: International Food Policy Research Institute, 2008 forthcoming).

CLIMATE CHANGE AWARENESS AND ADAPTATION

The study uses two separate models to examine the factors influencing farmers' decision to adapt to perceived climate changes. The models confirm that household wealth, represented by farm and nonfarm income and livestock ownership, increases the likelihood of climate change awareness and adaptation. Farmers are more likely to engage in soil conservation, use different crop varieties, and change planting dates as farm income increases. Nonfarm income increases the likelihood of planting trees, changing planting dates, and using irrigation as adaptation options. Additionally, ownership of livestock significantly increases the probability of adapting to climate change.

Household characteristics—including the level of education, sex and age of the household head, and household size—are found to increase the probability of adaptation. A higher level of education is associated with greater access to information on climate change, improved technologies, and higher productivity. The results show that education increases climate change awareness and the likelihood of soil conservation and changing planting dates as an adaptation method. Male-headed households are found to be more likely to adapt to climate change. Given that men do much of the agricultural work in Ethiopia, they are more likely to obtain information about and have access to new technologies, and to take greater risks than female-headed households. The age of the household head, which captures farming experience, also influences awareness of, and adaptation to, climate change. Similarly, larger households are more likely to adapt to perceived climate change, probably because they are often associated with a higher labor endowments.

Access to rural services, including credit, crop and livestock extension, and information about climate change, also increases the use of adaptation options. Availability of credit eases cash constraints and allows farmers to purchase inputs such as fertilizer, improved crop varieties, and irrigation facilities. The study confirms a positive relationship between the level of adoption and the availability of credit. In particular, access to credit increases the likelihood that farmers will employ soil conservation methods, change planting dates, and irrigate. Access to crop and livestock extension significantly increases the likelihood of adaptation, particularly the probability of planting trees. Information about climate change and appropriate adaptation methods also increases the likelihood of adaptation.

Social networks play distinct roles in adoption of agricultural technologies: they act as conduits for financial transfers that may ease farmers' credit constraints, provide information about new technologies, and facilitate cooperation among farmers to allow the costs and benefits of adaptation to be shared. The study confirms that social networks—measured by the number of relatives living in the area and access to farmer-to-farmer extension—increase awareness of climate change. In addition, the study finds that access to farmer-to-farmer extension increases the likelihood of adaptation to climate change.

Farm location, local temperature, and the amount of precipitation also influence farmers' adaptation to climate change in the Nile Basin of Ethiopia. As expected, farmers living in different agroecological settings employ different adaptation methods. For instance, lowland farmers are more likely to conserve soil but less likely to use different crop varieties, to plant tress, or to irrigate compared with midland farmers. Further, while highland farmers are more likely to perceive changes in climate than midland farmers, those in the highlands are less likely to plant trees.

Farms in areas with higher annual mean temperatures over the survey period were more likely to adapt to climate change. Higher annual mean temperature increases the likelihood of employing soil conservation methods, using different crop varieties, changing planting dates, and irrigating. Similarly, lower levels of precipitation over the survey period increased the likelihood of adopting adaptation techniques—specifically, using soil conservation methods, changing crop varieties, changing planting dates, and irrigating. These results suggest that as the temperature rises and conditions become drier, farmers employ methods to preserve soil moisture to ensure that their crops survive.

POLICY IMPLICATIONS

Findings indicate that policymakers and other stakeholders would do well to raise awareness of climate change, facilitate the availability of credit, invest in technologies, create additional opportunities for off-farm employment, invest in research on the use of new crop varieties and livestock species that are more suited to drier conditions, encourage informal social networks, and invest in irrigation. Such measures would help farmers in the Nile Basin of Ethiopia to moderate the adverse consequences of climate change, while maintaining their livelihoods and food security.

FOR FURTHER READING

Deressa, T. T., R. M. Hassan, C. Ringler, T. Alemu, and M. Yesuf, *Analysis of the Determinants of Farmers' Choice of Adaptation Methods and Perceptions of Climate Change in the Nile Basin of Ethiopia*, IFPRI Discussion Paper No. 798 (Washington, DC: International Food Policy Research Institute, 2008).

T. T. Deressa (tderessa@yahoo.com) is a Ph.D. student at the Centre for Environmental Economics and Policy for Africa (CEEPA), University of Pretoria. **R. M. Hassan** (rashid.hassan@up.ac.za) is director of CEEPA. **C. Ringler** (c.ringler@cgiar.org) is a senior research fellow at the International Food Policy Research Institute (IFPRI), Washington, DC. **T. Alemu** (tekie_a@econ.aau.edu.et) is a research fellow at the Ethiopian Development Research Institute (EDRI). **M. Yesuf** (mahmudyesuf@yahoo.com) is a senior research fellow at EDRI and an assistant professor at Addis Ababa University.

This series of IFPRI Research Briefs is based on research supported by the Federal Ministry for Economic Cooperation and Development, Germany, under the project "Food and Water Security under Global Change: Developing Adaptive Capacity with a Focus on Rural Africa," which forms part of the CGIAR Challenge Program on Water and Food. Through collaboration with the Center for Environmental Economics and Policy in Africa, the Ethiopian Development Research Institute, the Ethiopian Economics Association, and the University of Hamburg, the project aims to provide policymakers and stakeholders in Ethiopia and South Africa with tools to better understand and analyze the consequences of global change—in particular climate change—and to form policy decisions that facilitate adaptation in these countries and beyond.

Financial Contributors and Partners

IFPRI's research, capacity strengthening, and communications work is made possible by its financial contributors and partners. IFPRI receives its principal funding from governments, private foundations, and international and regional organizations, most of which are members of the Consultative Group on International Agricultural Research (CGIAR). IFPRI gratefully acknowledges the generous unrestricted funding from Australia, Canada, China, Finland, France, Germany, India, Ireland, Italy, Japan, Netherlands, Norway, South Africa, Sweden, Switzerland, United Kingdom, United States, and World Bank.

Printed on alternative-fiber paper manufactured from agriculturally sustainable resources that are processed chlorine-free (PCF).

INTERNATIONAL FOOD
POLICY RESEARCH INSTITUTE
sustainable solutions for ending hunger and poverty
IFPRI®
Supported by the CGIAR

CEEPA
Center for Environmental Economics
and Policy in Africa

IFPRI Research Brief 15–10

HOW CAN AFRICAN AGRICULTURE ADAPT TO CLIMATE CHANGE? INSIGHTS FROM ETHIOPIA AND SOUTH AFRICA

Climate Variability and Maize Yield in South Africa

Wisdom Akpalu, Rashid M. Hassan, and Claudia Ringler

Maize is the primary food staple in southern Africa, and 50 percent of the total maize output in the area is produced in South Africa, where maize constitutes approximately 70 percent of grain production and covers 60 percent of the country's cropping area. Climate change could have a significant impact on South African maize production. The scientific community has established that the temperature in South Africa increased significantly between 1960 and 2003 (by 0.13 degrees Celsius), and further temperature increases and changes in the quantity and pattern of rainfall are expected despite any attempts by the international community to reduce greenhouse gas emissions. Although the maize plant is quite hardy and adaptable to harsh conditions, warmer temperatures and lower levels of precipitation could have detrimental effects on yields, thereby increasing food insecurity in the region.

This brief is based on a paper that uses household survey data to explore the direct impact of climate variability, measured by changes in temperature and precipitation, on maize yields in the Limpopo Basin of South Africa.

MAIZE PRODUCTION IN SOUTH AFRICA

Sampled farms in the Limpopo Basin of South Africa produced 1,237 kilograms per hectare of maize using an average of 449 hours of labor per hectare, 26 kilograms of seed per hectare, and 159 kilograms of fertilizer per hectare during the 2004/05 growing season. A majority of the farms were rainfed; only 7 of the 25 farms in the survey sample (28 percent) used supplementary irrigation during the 2004/05 growing period (Table 1). Temperature and precipitation data were obtained from weather services in South Africa and were matched with farms within the neighborhood of each climate station. The mean temperature for the months of the 2004/05 farming season was 21.4 degrees Celsius, and the mean monthly precipitation was 71 millimeters (Table 1).

Table 1 Maize Production in the Limpopo Basin of South Africa, October 2004–May 2005

Indicator	Mean
Yield (kg/ha)	1,237
Labor (hrs/ha)	449
Seed (kg/ha)	26
Fertilizer (kg/ha)	159
Mean temperature (°C/month, October–May)	21
Mean precipitation (mm/month, October–May)	71
Irrigation (=1)	0.28

THE IMPACT OF CLIMATE VARIABILITY AND CHANGE ON MAIZE PRODUCTION IN SOUTH AFRICA

Under the study on which this brief is based, mathematical models were applied to estimate the direct impact of climate variability on maize yields. As was expected, an increase in production inputs—including labor, seed, fertilizer, and irrigation—raises maize yields substantially. Consistent with previous findings on the impact of climate change and crops in South Africa, the results suggest that a change in the amount of precipitation is the most important driver of maize yields. A 10 percent reduction in mean precipitation reduces the mean maize yield by approximately 4 percent. Correspondingly, an increase in mean precipitation increases mean maize yields; however, as rainfall continues to increase, the additional gain in maize yield begins to diminish (Figure 1). Also consistent with previous studies, the results suggest that changes in temperature affect maize yields. As the mean temperature increases from 21.4 to 21.6 degrees Celsius, the average maize yield increases by 0.4 percent. However, like increased precipitation, the gain in maize yields prompted by increased temperature begins to diminish as temperature increases further.

Figures 1 and 2 show that an increase in either precipitation or temperature from the 2004/05 mean values would increase maize yields at a decreasing rate. The combined effect of changes in temperature and rainfall on maize yields depends on the magnitude and direction of each of the changes. As predicted by climate models, the overall impact on yields of a marginal decrease in mean precipitation simultaneous with a marginal increase in mean temperature will be negative because the effect of reduced precipitation on maize yields is stronger than the effect of increased temperature. The figures also show that yields from

Figure 1 Diminishing Marginal Yield Benefit from Increasing Monthly Precipitation

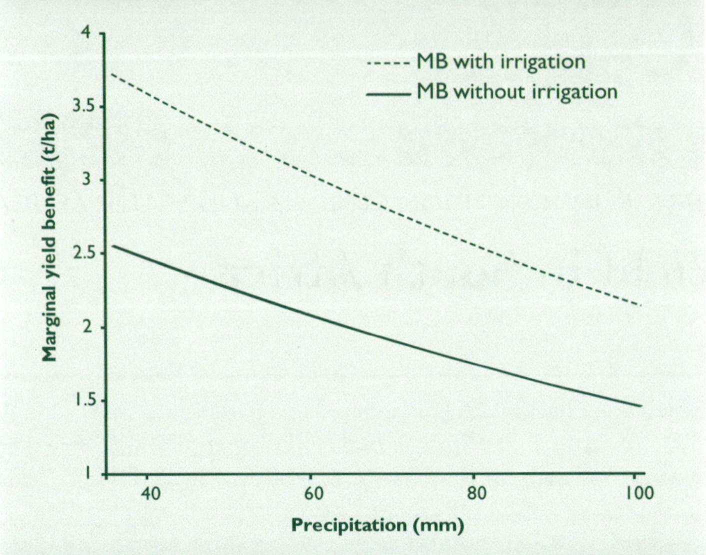

Figure 2 Diminishing Marginal Yield Benefit from Rising Temperature

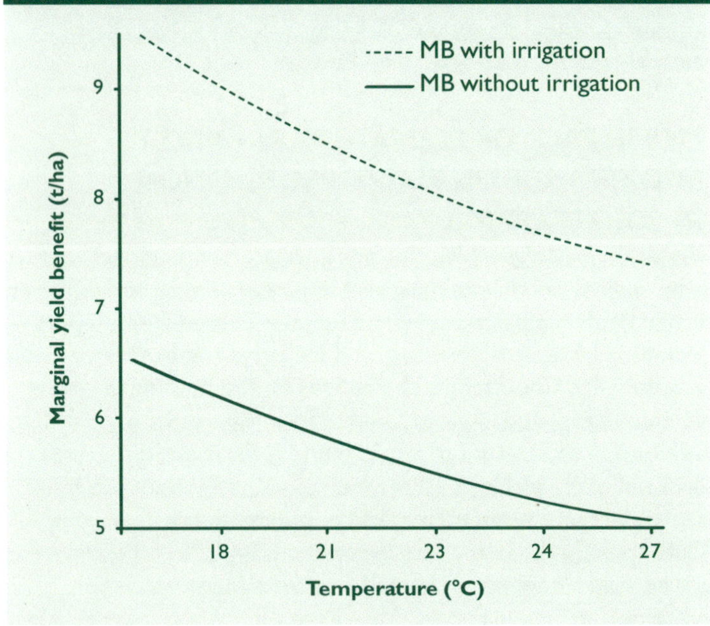

irrigated farms are higher than from non-irrigated farms, irrespective of the temperature and level of precipitation.

POLICY IMPLICATIONS

The scientific evidence shows that mean temperature in South Africa has increased and is expected to increase further in the future. At the same time, mean rainfall is expected to decrease by 5 to 10 percent, and rainfall variability is expected to increase over the next 50 years. The results of this study indicate that such effects would have a significant negative impact on maize yields and consequently pose a serious threat to food security in South Africa as well as other countries in the southern African region that depend on maize imports from South Africa.

The results also suggest that one way to mitigate potential yield loss due to climate change is to encourage irrigation. The findings show that irrigated farms had higher maize yields than did dryland farms; however, maize yields are determined more by the level of precipitation than by the presence of irrigation, indicating that irrigation practices partially mitigate the impact of decreased precipitation on yields.

Additional observations and data on temperature and precipitation at the farm level, rather than from nearby climate stations, would increase the robustness of these results. Nevertheless, while the study on which this brief is based could be improved with better data, this research provides an important starting point for further studies in South Africa and other developing countries on the impact of climate variability and climate change on crop yields and the resulting implications for food security.

FOR FURTHER READING

Akpalu, W., R. M. Hassan, and C. Ringler, *Climate Variability and Maize Yield in South Africa: Results from GME and MELE Methods*, IFPRI Discussion Paper No. 843 (Washington, DC, 2009).

W. Akpalu (akpaluw@farmingdale.edu) is an assistant professor at the State University of New York–Farmingdale. **R. M. Hassan** (rashid.hassan@up.ac.za) is the director of the Center for Environmental Economics and Policy in Africa (CEEPA) at the University of Pretoria, South Africa. **C. Ringler** (c.ringler@cgiar.org) is a senior research fellow in the Environment and Production Technology Division of the International Food Policy Research Institute.

This series of IFPRI Research Briefs is based on research supported by the Federal Ministry for Economic Cooperation and Development, Germany, under the project "Food and Water Security under Global Change: Developing Adaptive Capacity with a Focus on Rural Africa," which forms part of the CGIAR Challenge Program on Water and Food. Through collaboration with the Center for Environmental Economics and Policy in Africa, the Ethiopian Development Research Institute, the Ethiopian Economics Association, and the University of Hamburg, the project aims to provide policymakers and stakeholders in Ethiopia and South Africa with tools to better understand and analyze the consequences of global change—in particular climate change—and to form policy decisions that facilitate adaptation in these countries and beyond.

Financial Contributors and Partners

IFPRI's research, capacity strengthening, and communications work is made possible by its financial contributors and partners. IFPRI receives its principal funding from governments, private foundations, and international and regional organizations, most of which are members of the Consultative Group on International Agricultural Research (CGIAR). IFPRI gratefully acknowledges the generous unrestricted funding from Australia, Canada, China, Finland, France, Germany, India, Ireland, Italy, Japan, Netherlands, Norway, South Africa, Sweden, Switzerland, United Kingdom, United States, and World Bank.

Printed on alternative-fiber paper manufactured from agriculturally sustainable resources that are processed chlorine-free (PCF).

INTERNATIONAL FOOD POLICY RESEARCH INSTITUTE
sustainable solutions for ending hunger and poverty
IFPRI®
Supported by the CGIAR

EDRI
Ethiopia Development Research Institute

IFPRI Research Brief 15–11

HOW CAN AFRICAN AGRICULTURE ADAPT TO CLIMATE CHANGE? INSIGHTS FROM ETHIOPIA AND SOUTH AFRICA

The Impact of Climate Change and Adaptation on Food Production in Low-Income Countries

Evidence from the Nile Basin, Ethiopia

Mahmud Yesuf, Salvatore Di Falco, Temesgen Deressa, Claudia Ringler, and Gunnar Kohlin

Growing consensus in the scientific community indicates that higher temperatures and changing precipitation levels resulting from climate change will depress crop yields in many countries over the coming decades. This is particularly true in low-income countries, where adaptive capacity is low. Many African countries are particularly vulnerable to climate change because their economies largely depend on climate-sensitive agricultural production. This brief is based on a study that used household survey data to analyze the impact of climate change on food production in the Nile Basin of Ethiopia. The study also examined the factors influencing adaptation and the implications of various adaptation strategies for farm productivity.

CLIMATE CHANGE AND AGRICULTURAL PRODUCTION IN ETHIOPIA

With a population of more than 70 million people and a gross domestic product (GDP) of slightly more than US$10 billion, Ethiopia is one of the world's least developed countries. The agricultural sector, which is dominated by small-scale, subsistence farming, forms the foundation of the national economy and constitutes the primary source of livelihood for the overwhelming majority of the population. In 2003, the sector employed more than 80 percent of the labor force and contributed 45 percent of GDP and 85 percent of total export revenues. Ethiopian agriculture is almost exclusively dependent on rainfall, given that irrigated agriculture accounts for less than 1 percent of the country's total cultivated land. Thus, the amount and temporal distribution of rainfall and other climatic factors are key determinants of crop yields, and poor or excessive rainfall can induce food shortages and famine.

A recent mapping of vulnerability and poverty in Africa determined that, in terms of climate change, Ethiopia was one of the most vulnerable countries given its low adaptive capacity. Rainfall variability and associated drought have been major causes of the country's food shortages and famine. Nationally, the link between drought and crop production is widely recognized, but little is known about how climate change affects crop yields and what strategies households are using to adapt. Furthermore, few studies have analyzed the factors governing farmers' decisions to adapt to climate change and the impact of those decisions on yields. This information is particularly important for the design of effective adaptation strategies for coping with the negative impacts of climate change.

CLIMATE CHANGE AND ADAPTATION IN THE NILE BASIN OF ETHIOPIA

The survey underlying the study on which this brief is based was designed to capture farmers' perceptions and understanding of climate change, as well as their approaches to adaptation. Data show that 68 percent of farmers perceived that mean temperatures had increased over the previous 20 years, whereas 4 percent perceived they had decreased, and 28 percent perceived that there had been no change. In terms of mean annual rainfall over the same timeframe, 62 percent of farmers reported a decrease, 18 percent reported an increase, and 20 percent reported no change. Overall, increased temperature and declining precipitation were the predominant perceptions in the study sites.

In response to perceived long-term changes, farm households implemented a number of adaptation measures, including changing crop varieties, adopting soil and water conservation measures, harvesting water, planting trees, and changing planting and harvesting periods. The remaining adaptation measures, which accounted for less than 5 percent of all measures, were nonyield related and included migration and a shift in farming practices from crop production to livestock herding or other sectors. However, about 58 percent of farmers took no action to adapt to long-term shifts in temperatures, and 42 percent took no action to respond to long-term shifts in precipitation. More than 90 percent of those respondents who took no action to adapt cited lack of information and shortages of labor, land, and money as the major reasons. In fact, lack of information was the predominant reason cited by 40–50 percent of households.

DETERMINANTS OF ADAPTATION

Results suggest that information about future changes in climate and access to formal and informal institutions strongly govern household decisions about adaptation. Households with access to formal agricultural extension, farmer-to-farmer extension, credit, and information about future climate change are more likely to adjust their farming practices in response to climate change. Likewise, households in areas that experienced higher rainfall than average in the Belg (fall) season were also more likely to adopt adaptation strategies compared with households in areas receiving less rainfall. Nevertheless, households with higher than average rainfall during the key Mehere (summer) rainfall season were not more likely to adapt to climate change.

Significant differences were also observed across the country's various agroecological zones when it came to the likelihood that households would undertake measures to adapt to climate change: households in the highlands (Dega) and midlands (WeynaDega) were less likely to adopt adaptation measures compared with households in the lowlands (Kolla). Significant differences in responses were also observed based on household size and age and literacy levels of household heads. Larger households and those whose heads were older and more literate were more likely to adopt adaptation measures, indicating the importance of available labor on the one hand and experience and access to information on the other.

THE IMPACT OF CLIMATE CHANGE AND ADAPTATION ON FOOD PRODUCTION

Although the survey returned information on a total of 48 annual crops grown in the Nile Basin of Ethiopia, the first 5 major annual crops (teff, maize, wheat, barley, and beans) cover 65 percent of plots. The estimation of the production function that follows is limited to these crops.

Results show that farmers who adopted measures to adapt to climate change had higher levels of food production than those who did not: households that adopted adaptation measures tended to produce about 95–300 kilograms more food per hectare than those who chose not to adapt, which accounts for a 10–29 percent difference in output. In other words, adaptation measures substantially mitigated the effect of climate change on crop yields.

Farm-level climatic variation is a significant factor in explaining fluctuations in food production across farm households. Variations in household yield levels could not be explained by temperature. Variations in precipitation during the Mehere season did explain yield differences, although the effects of such differences were nonlinear. An increase in both Belg and Mehere rainfall seemed to increase food production (controlling for agroecological and other major factors of production), but too much or too little rainfall during these seasons appeared to have a negative effect on food production in the study sites.

As expected, the use of improved seeds, fertilizers, manure, and additional labor tended to increase food production. Significant differences in yields were also observed across agroecological zones, with the highlands (Dega) producing the most food per hectare, followed by the lowlands (Kolla), and finally the midlands (WeynaDega).

CONCLUDING REMARKS

The above results indicate that farmers' decisions to adopt yield-enhancing adaptation strategies are influenced by informal and formal institutional support, the availability of information on future climate changes, the amount of rainfall during the Belg season, and the agroecological setting, as well as household-specific characteristics of size and age and literacy levels of the household head. This suggests that farmers need appropriate and timely information on predicted changes in climate to empower them to take appropriate steps to adjust their farming practices. Moreover, given that access to credit markets and government and farmer-to-farmer extension services was shown to facilitate adaptation, more effort should be made to extend these services to farmers in poor communities.

Averting the negative effects of climate change and achieving food security have become major priorities for development agencies, policymakers, and related stakeholders. Given that adaptation measures have a positive effect on crop yields, the adoption of yield-related adaptation strategies could significantly support these goals. Consequently, adaptation not only enables farmers to cope with the adverse effects of climate change and variability, but also increases the agricultural productivity of poor farm households.

FOR FURTHER READING

Yesuf, M., S. Di Falco, T. Deressa, C. Ringler, and G. Kohlin, *The Impact of Climate Change and Adaptation on Food Production in Low-Income Countries: Evidence from the Nile Basin, Ethiopia*, IFPRI Discussion Paper No. 828 (Washington, DC: International Food Policy Research Institute, 2008).

M. Yesuf (mahmudyesuf@yahoo.com) is a fellow at the Ethiopian Development Research Institute and Addis Ababa University. **S. Di Falco** (s.difalco@uea.ac.uk) is a lecturer in Applied Economics at the University of Kent, U.K. **T. Deressa** (ttderessa@yahoo.com) is a PhD student at the Center for Environmental Economics and Policy in Africa (CEEPA) at the University of Pretoria, South Africa. **C. Ringler** (c.ringler@cgiar.org) is a senior research fellow in the Environment and Production Technology Division of the International Food Policy Research Institute. **G. Kohlin** (gunnar.kohlin@economics.gu.se) is an associate professor in Economics at Gothenburg University, Sweden.

This series of IFPRI Research Briefs is based on research supported by the Federal Ministry for Economic Cooperation and Development, Germany, under the project "Food and Water Security under Global Change: Developing Adaptive Capacity with a Focus on Rural Africa," which forms part of the CGIAR Challenge Program on Water and Food. Through collaboration with the Center for Environmental Economics and Policy in Africa, the Ethiopian Development Research Institute, the Ethiopian Economics Association, and the University of Hamburg, the project aims to provide policymakers and stakeholders in Ethiopia and South Africa with tools to better understand and analyze the consequences of global change—in particular climate change—and to form policy decisions that facilitate adaptation in these countries and beyond.

Financial Contributors and Partners

IFPRI's research, capacity strengthening, and communications work is made possible by its financial contributors and partners. IFPRI receives its principal funding from governments, private foundations, and international and regional organizations, most of which are members of the Consultative Group on International Agricultural Research (CGIAR). IFPRI gratefully acknowledges the generous unrestricted funding from Australia, Canada, China, Finland, France, Germany, India, Ireland, Italy, Japan, Netherlands, Norway, South Africa, Sweden, Switzerland, United Kingdom, United States, and World Bank.

Printed on alternative-fiber paper manufactured from agriculturally sustainable resources that are processed chlorine-free (PCF).

INTERNATIONAL FOOD
POLICY RESEARCH INSTITUTE
sustainable solutions for ending hunger and poverty
IFPRI®
Supported by the CGIAR

IFPRI Research Brief 15–12

HOW CAN AFRICAN AGRICULTURE ADAPT TO CLIMATE CHANGE? INSIGHTS FROM ETHIOPIA AND SOUTH AFRICA

Impacts of Considering Climate Variability on Investment Decisions in Ethiopia

Paul J. Block, Kenneth Strzepek, Mark Rosegrant, and Xinshen Diao

Numerous studies indicate that agricultural production is sensitive to climate variability, and lack of infrastructure in developing countries increases vulnerability to extreme climate events. In Ethiopia, the historical climate record indicates frequent droughts and floods, which can devastate agricultural production and existing infrastructure. Too much precipitation can flood crops, rot or suffocate roots, and wash out roads, creating similar economic conditions to those resulting from drought. With 85 percent of the population living in rural areas, and most people depending on rainfed agriculture, Ethiopia's social and economic welfare depends heavily on climatic conditions.

This brief is based on a paper that uses an economywide, multi-sector, and multi-regional model to assess the impact of climate variability on the outcomes of prospective investment strategies for Ethiopia, as well as on the country's gross domestic product growth rates and poverty rates.

INVESTMENT STRATEGIES AND SIMULATIONS

The model brings agricultural supply, demand, and market opportunity issues together to assess different investment strategies. The analysis gives a broad picture of agricultural growth and poverty reduction and reveals some important economic linkages among agricultural sectors and between demand and supply, exports and domestic markets, and production and farmer income. Additionally, the analysis reveals the complexity of economic linkages and trade-offs among different investment goals. Model results are limited to four economic indicators: the poverty rate and growth rates for gross domestic product (GDP), agricultural GDP, and nonagricultural GDP.

Two major investment strategies are simulated in the model: investments in irrigation, which focus on providing sufficient water to crops, and investments in road construction and maintenance, which focus on ensuring that farmers are able to transport their agricultural products to local markets and for export. A third simulation simultaneously incorporates investments in both irrigation and roads.

For irrigation, the model simulates the effects of investment by incorporating the country's Irrigation Development Program under the Ethiopian Ministry of Water Resources' Water Sector Development Plan (WSDP 2002). Approximately 200,000 hectares of crop area are currently being irrigated in Ethiopia, accounting for just over 2 percent of all cropland. The new program details the addition of 274,000 hectares of irrigated cropland, more than doubling the current irrigated area. For roads, the model simulates the effects of investment based on Ethiopia's 10-year Road Sector Development Program (RSDP 1997–2007). Prior to the implementation of RSDP, Ethiopia's road network consisted of approximately 3,800 kilometers of paved roads and 29,000 kilometers of unpaved roads. The first half of the program (1997–2002) focused on rehabilitating the existing roadways, while the second half aimed both to continue rehabilitation and to increase the network of all types of roads by 5,000 to 8,000 kilometers.

The results of the three simulations are compared against a baseline simulation, which predicts future conditions if current practices remain unchanged—that is, with no infrastructure investments or major policy changes. The baseline scenario stays within the confines of historical growth rates under mean climate conditions for each year, resulting in a smooth growth trend.

THE EFFECTS OF MODELING CLIMATE VARIABILITY

The economic and poverty-reducing effects of investing in irrigation and roads under average climate conditions for a single model run (a deterministic model) are compared with the effects of variable climate conditions for numerous model runs (a stochastic model), including year-to-year changes and extreme climate events. In the deterministic model, the effects of drought and flood are ignored through the use of average climate parameters, whereas the effects of both drought and flood are included in the stochastic model.

The analysis suggests that climate variability results in lower economic welfare; in particular, it highlights the difficulty of returning to a state of constant growth after a shock. Average climate parameters, as used in the deterministic model, can underestimate the negative effects of climate variability and do not clearly represent the difficulties of recovering from extreme climate events. The inclusion of drought and flood impacts in this study results not only in expected declines in agricultural yields, but also in potential damage to roads and infrastructure, which exacerbates declines in agricultural production, trade, and other nonagricultural activities. Failure to include these factors may well result in misleading insights and an overestimation of the welfare benefits to the economy (Table 1). Stochastic modeling, including climate variability, thus appears to be essential and warranted when modeling investment in sectors that are responsive to climate extremes. Including climate variability in the model offers more prudent predictions based on realistic ranges of economic values from which Ethiopian planners may make strategic decisions.

The simulations of increased investment in irrigated agriculture, roads, and a combination of the two showed improved economic outcomes compared with the baseline case. The results of the simulations indicate that benefits from investments in irrigation and roads are manifested differently; investments in irrigation provide benefits within the agricultural sector, while investments in roads predominantly benefit the nonagricultural sector, but they allow for some agricultural feedbacks. The results also show that investments in irrigation are slightly more effective in boosting the Ethiopian economy and reducing poverty than are investments in roads, in part because additional irrigation ameliorates the negative effects of drought and therefore has a particularly strong impact on production and farm income. Because drought has a persistent impact on income and food security, preventing or reducing the severity of drought has long-term benefits. As expected, the welfare of the country is best under the combined investment strategy.

Finally, considering climate variability when predicting the outcome of investment decisions can also provide some sense of probabilistic risk. No investment strategy can ensure economic success, so planners need to know the probability that the economy will fall within some acceptable range of economic values for a given investment strategy.

FOR FURTHER READING

Block, P. J., K. Strzepek, M. Rosegrant, and X. Diao, *Impacts of Considering Climate Variability on Investment Decisions in Ethiopia*, Environment and Production Technology Division Discussion Paper No. 150 (Washington, DC: IFPRI, 2006).

Table 1 Simulated effects of investments on poverty and GDP growth rates

Simulation	GDP growth rate: Average climate conditions for a single model run (deterministic model)	GDP growth rate: Range for variable climate conditions for numerous model runs (stochastic model)	Poverty rate 2015: Average climate conditions for a single model run (deterministic model)	Poverty rate 2015: Range for variable climate conditions for numerous model runs (stochastic model)
Baseline	2.82	2.34–3.03	42.98	40.59–52.22
Irrigation	3.68	2.25–3.22	39.27	42.70–61.36
Roads	3.58	2.00–3.08	39.82	43.37–62.96
Combination of irrigation and roads	4.40	2.95–3.92	36.15	39.77–58.77

SOURCE: Block, P. J., Kenneth Strzepek, Mark Rosegrant, and Xinshen Diao, *Impacts of Considering Climate Variability on Investment Decisions in Ethiopia*, Environment and Production Technology Division Discussion Paper No. 150 (Washington, DC: IFPRI, 2006).

P. J. Block (pblock@iri.columbia.edu) is a postdoctoral research scientist at the International Research Institute for Climate and Society of Columbia University, New York. **K. Strzepek** (Kenneth.Strzepek@Colorado.edu) is professor of civil, environmental, and architectural engineering at the University of Colorado, Boulder. **M. Rosegrant** (m.rosegrant@cgiar.org) is director of the Environment and Production Technology Division of the International Food Policy Research Institute (IFPRI). **X. Diao** (x.diao@cgiar.org) is a senior research fellow in the Development Strategy and Governance Division at IFPRI.

This series of IFPRI Research Briefs is based on research supported by the Federal Ministry for Economic Cooperation and Development, Germany, under the project "Food and Water Security under Global Change: Developing Adaptive Capacity with a Focus on Rural Africa," which forms part of the CGIAR Challenge Program on Water and Food. Through collaboration with the Center for Environmental Economics and Policy in Africa, the Ethiopian Development Research Institute, the Ethiopian Economics Association, and the University of Hamburg, the project aims to provide policymakers and stakeholders in Ethiopia and South Africa with tools to better understand and analyze the consequences of global change—in particular climate change—and to form policy decisions that facilitate adaptation in these countries and beyond.

Financial Contributors and Partners

IFPRI's research, capacity strengthening, and communications work is made possible by its financial contributors and partners. IFPRI receives its principal funding from governments, private foundations, and international and regional organizations, most of which are members of the Consultative Group on International Agricultural Research (CGIAR). IFPRI gratefully acknowledges the generous unrestricted funding from Australia, Canada, China, Finland, France, Germany, India, Ireland, Italy, Japan, Netherlands, Norway, South Africa, Sweden, Switzerland, United Kingdom, United States, and World Bank.

Printed on alternative-fiber paper manufactured from agriculturally sustainable resources that are processed chlorine-free (PCF).

INTERNATIONAL FOOD POLICY RESEARCH INSTITUTE
sustainable solutions for ending hunger and poverty
IFPRI®
Supported by the CGIAR

IFPRI Research Brief **15–13**

HOW CAN AFRICAN AGRICULTURE ADAPT TO CLIMATE CHANGE? INSIGHTS FROM ETHIOPIA AND SOUTH AFRICA

Global Carbon Markets

Are There Opportunities for Sub-Saharan Africa?

Elizabeth Bryan, Wisdom Akpalu, Claudia Ringler, and Mahmud Yesuf

Human activities such as fossil fuel burning and deforestation have significantly increased the atmospheric concentration of greenhouse gases (GHG) leading to global climate change. Global climate change and its associated weather extremes pose considerable challenges worldwide, and mitigating the adverse impacts of climate change is a high priority for the international community.

To reduce global emissions and curb the threat of climate change, many countries are participating in carbon trading. Carbon trading includes allowance-based agreements that impose national caps on emissions and allow participating countries to engage in emission trading as well as project-based transactions (for example, through the CDM or Clean Development Mechanism). The CDM allows industrialized countries with greenhouse gas reduction commitments to invest in emission-reducing projects in developing countries as an alternative to generally more costly emission reductions in their own countries. Funds made available by the CDM for carbon offsets provide an opportunity for cash-strapped developing countries to fund much needed adaptation measures. The potential annual value stream for Sub-Saharan Africa from mitigating GHG emissions is estimated to be US$4.8 billion at carbon prices of US$0–20/t$CO_2$e. Moreover, agricultural mitigation measures, including soil and water conservation and agroforestry practices, also enhance ecosystem functioning, providing resilience against droughts, pests, and climate-related shocks.

Yet the potential for Africa to contribute to global reductions in GHG emissions is quite substantial. Estimates suggest Africa could potentially contribute to GHG reductions of 265 MtCO_2e (million tons of carbon dioxide or equivalent) per year at carbon prices of up to US$20 through agricultural measures and 1,925 MtCO_2e/yr at carbon prices of up to US$100/t$CO_2$e by 2030 through changes in the forestry sector. These amounts constitute 17 and 14 percent, respectively, of the global total potential for mitigation in these sectors. However, countries in Sub-Saharan Africa are marginalized in global carbon markets. Sub-Saharan Africa's share of the CDM market is nine times smaller than its global share of GHG emissions, including emissions from land use and land-use change.

This brief is based on a paper that examines Sub-Saharan Africa's current involvement in carbon markets, potential for GHG emission reductions, constraints to further participation in carbon markets, and opportunities for expanding Sub-Saharan Africa's market share.

SUB-SAHARAN AFRICA'S MARKET SHARE AND POTENTIAL

As the largest project-based market aimed at developing countries, the CDM provides the largest outlet for carbon offset projects in Sub-Saharan Africa. As of October 2008, Sub-Saharan Africa accounted for only 1.4 percent of all registered CDM projects—only 17 out of 1,186 projects—and most of these projects (14 out of 17) were located in just one country, South Africa. Thus, African projects still represent a small fraction of the entire CDM market. China dominates the CDM market with about 73 percent of volumes transacted (in 2007).

While Sub-Saharan Africa's contribution to global emissions is relatively small—5 percent of the global total—there is significant potential for the region to contribute to climate change mitigation, particularly in the forestry and agriculture sectors, which together accounted for 73 percent of emissions from the region (and 13 percent of the global total emissions from these sectors). Moreover, Africa's emissions from agriculture and land-use change and deforestation are

Table 1 Estimated Economic Mitigation Potential by Management Practice and Region

	Economic Mitigation Potential by 2030 at US$0–20/t$CO_2$e (Mt$CO_2$e/yr)					
	Cropland management	Grazing land management	Restoration of organic soils	Restoration of degraded land	Other practices	Total
East Africa	28	27	25	13	15	109
Central Africa	13	12	11	6	7	49
North Africa	6	6	6	3	3	25
South Africa	6	5	5	3	3	22
West Africa	16	15	14	7	8	60
Total	69 (26%)	65 (25%)	61 (23%)	33 (12%)	37 (14%)	265 (100%)

SOURCE: Smith et al. 2008.

expected to grow in the future due to projected intensification of agricultural production and the expansion of unexploited areas.

The mitigation potential from agricultural production is greatest in East, West, and Central Africa, with mitigation potentials of 109, 60, and 49 $MtCO_2e$/yr, respectively, at prices of US$0–20/$tCO_2e$ (see Table 1). The agricultural practices that appear to be the most promising include cropland management, grazing/land management, and restoration of organic soils.

Moreover, Africa contributes 18 percent of the total global GHG emissions from land use, land-use change, and forestry. As such, soil carbon sequestration, fire management, and avoided deforestation offer additional opportunities for mitigating GHG emissions and promoting sustainability in Africa. Africa has the potential to mitigate 1,160 MtCO2e/yr from avoided deforestation by 2030, 29 percent of the global total, as well as 665 MtCO2e/yr from afforestation and 100 MtCO2e/yr from forest management at carbon prices of US$0–100/tCO2e.

CONSTRAINTS TO SUB-SAHARAN AFRICA'S PARTICIPATION IN GLOBAL CARBON MARKETS

While Sub-Saharan Africa could contribute considerably to global reductions in GHG emissions, numerous barriers would have to be overcome. For instance, to be considered eligible to engage in carbon trading under the CDM, a clear baseline for a project must be established, and it must be demonstrated that emission reductions would not have occurred in the absence of the project (additionality rule). For many developing countries, lack of technical training and support on setting benchmarks, as well as poor availability and quality of data, are major obstacles to defining an adequate baseline and demonstrating additionality.

Africa's participation in the CDM is also constrained by high transaction costs. The costs of carbon projects include the cost of providing information about carbon benefits to potential buyers, communicating with project partners, and ensuring parties fulfill their contractual obligations. Measurement and monitoring costs are also often considerable. Likewise, the costs of negotiating land-use decisions for carbon projects involving large numbers of geographically dispersed people with different land-use objectives can be prohibitive.

In addition, the CDM targets energy and power sources, overlooking soil carbon sequestration and avoided deforestation projects, which are highly important for climate change mitigation in many African countries. The exclusion of these activities limits CDM participation by African countries and hinders their mitigation opportunities.

OPPORTUNITIES FOR INTEGRATING SUB-SAHARAN AFRICA INTO THE GLOBAL CARBON MARKET

There are several opportunities for further integrating Sub-Saharan African and other developing countries into global carbon markets. Simplifying the CDM rules for determining baselines, monitoring carbon emissions, and enforcing offsets and broadening the range of eligible projects to include avoided deforestation and soil carbon sequestration would facilitate the participation of Sub-Saharan African countries. These countries should also explore opportunities to increase participation in voluntary carbon markets. In order to take full advantage of the opportunities provided by carbon markets, Sub-Saharan African countries will also need to strengthen their institutional capacity and engage both private and public sectors in project development and implementation. International advisory services could be established to assist potential investors, project designers and managers, national policymakers, and leaders of local organizations and federations in negotiating deals and complying with measurement and monitoring requirements.

Policymakers should take care to ensure that the needs of the poor are taken into consideration. Reducing the transaction costs associated with small-scale carbon offset projects would allow the poor within these countries to benefit from carbon trading. Working with intermediary organizations that are accountable to local producers, building community-management capacity, strengthening property rights, and improving regulation of offset projects would also help ensure that social and environmental goals are met and that the poor benefit from the carbon trading system. Thus, expanding pro-poor mitigation through linking Sub-Saharan Africa to global carbon markets is both feasible and desirable for the region in terms of conserving its natural resources, contributing to the good of the global environment, and generating income to finance its development activities.

FOR FURTHER READING

Bryan, E., W. Akpalu, C. Ringler, and M. Yesuf, 2008. *Global Carbon Markets: Are There Opportunities for Sub-Saharan Africa?* IFPRI Discussion Paper (Washington, DC: International Food Policy Research Institute, 2008 forthcoming).

Smith, P., D. Martino, Z. Cai, D. Gwary, H. Janzen, P. Kumar, B. McCarl, S. Ogle, F. O'Mara, C. Rice, B. Scholes, O. Sirotenko, M. Howden, T. McAllister, G. Pan, V. Romanenkov, U. Schneider, S. Towprayoon, M. Wattenbach, and J. Smith, "Greenhouse-Gas Mitigation in Agriculture," *Philosophical Transactions of the Royal Society* B 363 (2008).

E. Bryan (e.bryan@cgiar.org) is a senior research assistant in the Environment and Production Technology Division (EPTD) of the International Food Policy Research Institute. **W. Akpalu** (akpaluw@farmingdale.edu) is a professor at the State University of New York-Farmingdale. **C. Ringler** (c.ringler@cgiar.org) is a is a senior research fellow with EPTD at IFPRI. **M. Yesuf** (mahmudyesuf@yahoo.com) is a fellow at the Ethiopian Development Research Institute and Addis Ababa University.

This series of IFPRI Research Briefs is based on research supported by the Federal Ministry for Economic Cooperation and Development, Germany, under the project "Food and Water Security under Global Change: Developing Adaptive Capacity with a Focus on Rural Africa," which forms part of the CGIAR Challenge Program on Water and Food. Through collaboration with the Center for Environmental Economics and Policy in Africa, the Ethiopian Development Research Institute, the Ethiopian Economics Association, and the University of Hamburg, the project aims to provide policymakers and stakeholders in Ethiopia and South Africa with tools to better understand and analyze the consequences of global change—in particular climate change—and to form policy decisions that facilitate adaptation in these countries and beyond.

Financial Contributors and Partners

IFPRI's research, capacity strengthening, and communications work is made possible by its financial contributors and partners. IFPRI receives its principal funding from governments, private foundations, and international and regional organizations, most of which are members of the Consultative Group on International Agricultural Research (CGIAR). IFPRI gratefully acknowledges the generous unrestricted funding from Australia, Canada, China, Finland, France, Germany, India, Ireland, Italy, Japan, Netherlands, Norway, South Africa, Sweden, Switzerland, United Kingdom, United States, and World Bank.

Printed on alternative-fiber paper manufactured from agriculturally sustainable resources that are processed chlorine-free (PCF).

INTERNATIONAL FOOD POLICY RESEARCH INSTITUTE
sustainable solutions for ending hunger and poverty
IFPRI®
Supported by the CGIAR

CGIAR Challenge Program on WATER & FOOD

IFPRI Research Brief 15–14

HOW CAN AFRICAN AGRICULTURE ADAPT TO CLIMATE CHANGE? INSIGHTS FROM ETHIOPIA AND SOUTH AFRICA

Integrated Management of the Blue Nile Basin in Ethiopia under Climate Variability and Climate Change

Hydropower and Irrigation Modeling

Paul J. Block, Kenneth Strzepek, and Balaji Rajagopalan

Ethiopia possesses abundant water resources and hydropower potential, yet less than 5 percent of irrigable land in the Blue Nile basin has been developed for food production, and more than 80 percent of Ethiopians lack access to electricity. Consequently, the Ethiopian government is pursuing plans to develop hydropower and irrigation along the Blue Nile River in an effort to tap into this underused potential.

Although approximately 84 percent of the inflow to Lake Nasser at Aswan, Egypt, initiates from Ethiopia through the Blue Nile and Atbara Rivers, Ethiopia has limited rights to use these resources. Egypt and Sudan, through the Agreement of 1959, are allotted 55.5 and 18.5 billion cubic meters each year, respectively, with no allotment to Ethiopia. In 1998 the Nile Basin Initiative was created to stimulate cooperation among all countries in the Nile basin and work toward amicable alternatives and solutions for water resources benefits.

This brief is based on a paper that analyzes potential hydropower generation and irrigation supply for four large-scale dams and reservoirs along the Blue Nile River within Ethiopia—Karadobi, Mabil, Mendaia, and Border—as proposed by the United States Bureau of Reclamation in 1964. The total installed hydropower capacity would be 5,570 megawatts of power, about 2.5 times the potential of the Aswan High Dam in Egypt. Irrigation associated with the Mendaia and Border reservoirs could expand by 250,000 hectares or 35 percent of the estimated total irrigable land in the Blue Nile basin.

The challenges of implementation, however, are not inconsequential. The proposed reservoirs not only raise financing, investment, political, and institutional challenges, but may also require many years to fill, which will affect downstream flows depending on variable climate and climate change conditions. Using a model for hydropower and irrigation analysis, the paper explores dam implementation viability under various policy options.

PLAUSIBLE FLOW RETENTION POLICIES AND CLIMATE SCENARIOS

The model simulates two different flow policies, both of which represent plausible scenarios for retaining water within Ethiopia (neither of which is acceptable under current agreements). The first allows for Ethiopia to retain 5 percent of the annual flow (the 5-percent policy) passing the Sudano-Ethiopian border. The second only allows Ethiopia to retain water in years within which the annual border flow exceeds 50 percent of historical flows (the 50-percent policy), and in that event the entire excess may be withheld. The model provides a benefit–cost analysis of the implementation of this project, including both hydropower generation and irrigation development for the two flow policies under three different climate scenarios. One scenario projects future climate variability based on historical data (1961–90), whereas the other two scenarios project potential climate variability based on increased frequency of El Niño or La Niña events due to climate change.

Costs and benefits of project implementation are assessed over a 100-year period, 2000–99, which includes a construction phase of seven years (2000–06) for the first dam and three years (2004–06) for the irrigation system before any benefits are realized. Additional dams are constructed in subsequent seven-year periods. The transient (filling) phase of the model begins in 2007 when water may first be impounded in the initial dam. Postconstruction benefits (beyond 2036) are assumed to be constant at the design level.

HISTORICAL AND CLIMATE CHANGE SCENARIO RESULTS

Table 1 presents benefit–cost ratios for the three potential climate conditions and two flow policies. The expected benefit–cost ratios for an increased frequency of La Niña events are approximately equal to those of the historical scenario under the 5-percent flow retention policy. Under the 50-percent policy, the La Niña benefit–cost ratios are slightly greater than the historical scenario, owing to generally wetter conditions.

In contrast, the expected benefit–cost ratios for an increased frequency of El Niño events produce noticeably lower benefit–cost ratios compared with the historical scenario. This outcome is a direct result of generally drier conditions, particularly during the transient phase, and clearly represents conditions under which construction of the hydropower and irrigation projects may not prove worthwhile. Moreover, the benefit–cost ratios for the increased frequency of El Niño events may well be an overestimation because the likelihood of achieving full benefits for irrigation and hydropower beyond the transient stage is small.

IRRIGATION VERSUS HYDROPOWER

In the historical and La Niña scenarios, hydropower and irrigation are almost always both maximized. Irrigation benefit–cost ratios are generally close to 1.0. For drier conditions, however, such as the El Niño scenario, the model reserves water for hydropower generation and forgoes crop irrigation in order to meet downstream flow requirements.

For the El Niño 5-percent flow policy, the number of hectares irrigated in the early years may be lower than for other climatic conditions given the generally drier conditions. However, the number of hectares irrigated may grow quickly thereafter, reaching the maximum level within a two- to four-year period. For the El Niño 50-percent flow policy, however, little or no irrigation may take place during the transient stage, which is not helpful for cropland management and planning and, consequently, contributes to benefit–cost ratios below 1.0. Understandably, a surge in crop yields or commodity prices associated with a decrease in demand for energy may reverse these trends. Other political decisions—related to national food security, for example—might also favor irrigation over the energy development strategy.

CONCLUSION

Climate change could play a major role in determining the success or failure of proposed large-scale hydropower and irrigation projects in Ethiopia's Blue Nile Basin. Climate change scenarios, represented by changes in the frequency of El Niño and La Niña events, indicate potential for small benefit–cost increases, but they also reflect the potential for notable decreases relative to historical climate conditions. Stochastic modeling of scenarios representing a doubling of the historical frequency of El Niño events indicates benefit–cost ratios as low as 1.0, with numerous runs producing potentially infeasible hydropower and irrigation projects due to a lack of timely water. Overall, the 5-percent flow policy appears to be more robust to modeled climate changes than the 50-percent flow policy. It consistently outperforms the 50-percent flow policy in drier conditions and is nearly on par with it in wetter conditions.

Although considerable effort has been devoted to creating as comprehensive and accurate a model as possible, the Blue Nile within Ethiopia remains largely ungauged, and some degree of uncertainty must be factored into the use of specific hydrologic and climatic conditions. Undoubtedly, site-specific testing and modern technology will alter earlier reservoir plans, possibly changing the potential or overall scope for hydropower and irrigation development. Nonetheless, the results of this study are thought to be representative of prospective future hydropower and irrigation development scenarios and at least give an indication of feasibility under varying conditions.

Coordinating plans with downstream riparian countries is vital to the success of the hydropower and irrigation development projects. Potential benefits of collaboration for the countries involved include increased energy and food production, regulated streamflow, increased water conservation through reduced evaporation losses, and redistributed water rights through a renegotiation of the 1959 Agreement.

FOR FURTHER READING

Block, P. J., K. Strzepek, and B. Rajagopalan, *Integrated Management of the Blue Nile Basin in Ethiopia: Hydropower and Irrigation Modeling*, IFPRI Discussion Paper No. 700 (Washington, DC: International Food Policy Research Institute, 2007).

Table 1 Benefit–cost ratios for two flow policies for historical and climate change scenarios

Type of flow policy	Historical conditions	Increased frequency (2x) of La Niña (wetter conditions)	Increased frequency (2x) of El Niño (drier conditions)
5-percent policy	1.48–1.72	1.49–1.76	1.43–1.66
50-percent policy	1.18–1.82	1.41–1.91	1.07–1.63

SOURCE: Block, P. J., K. Strzepek, and B. Rajagopalan, *Integrated Management of the Blue Nile Basin in Ethiopia: Hydropower and Irrigation Modeling*, IFPRI Discussion Paper No. 700 (Washington, DC: International Food Policy Research Institute, 2007).
NOTE: The interest rate is 10 percent.

P. J. Block (pblock@iri.columbia.edu) is a postdoctoral research scientist at the International Research Institute for Climate and Society of Columbia University, New York. **K. Strzepek** (Kenneth.Strzepek@Colorado.edu) is professor of civil, environmental, and architectural engineering at the University of Colorado, Boulder. **B. Rajagopalan** (rajagopalan.balaji@colorado.edu) is assistant professor of civil, environmental, and architectural engineering at the University of Colorado, Boulder.

This series of IFPRI Research Briefs is based on research supported by the Federal Ministry for Economic Cooperation and Development, Germany, under the project "Food and Water Security under Global Change: Developing Adaptive Capacity with a Focus on Rural Africa," which forms part of the CGIAR Challenge Program on Water and Food. Through collaboration with the Center for Environmental Economics and Policy in Africa, the Ethiopian Development Research Institute, the Ethiopian Economics Association, and the University of Hamburg, the project aims to provide policymakers and stakeholders in Ethiopia and South Africa with tools to better understand and analyze the consequences of global change—in particular climate change—and to form policy decisions that facilitate adaptation in these countries and beyond.

Financial Contributors and Partners

IFPRI's research, capacity strengthening, and communications work is made possible by its financial contributors and partners. IFPRI receives its principal funding from governments, private foundations, and international and regional organizations, most of which are members of the Consultative Group on International Agricultural Research (CGIAR). IFPRI gratefully acknowledges the generous unrestricted funding from Australia, Canada, China, Finland, France, Germany, India, Ireland, Italy, Japan, Netherlands, Norway, South Africa, Sweden, Switzerland, United Kingdom, United States, and World Bank.

Printed on alternative-fiber paper manufactured from agriculturally sustainable resources that are processed chlorine-free (PCF).

INTERNATIONAL FOOD POLICY RESEARCH INSTITUTE
sustainable solutions for ending hunger and poverty
IFPRI®
Supported by the CGIAR

UNIVERSITY OF HAMBURG

IFPRI Research Brief 15–15

HOW CAN AFRICAN AGRICULTURE ADAPT TO CLIMATE CHANGE? INSIGHTS FROM ETHIOPIA AND SOUTH AFRICA

Economywide Impacts of Climate Change on Agriculture in Sub-Saharan Africa

Alvaro Calzadilla , Tingju Zhu, Katrin Rehdanz, Richard S. J. Tol, and Claudia Ringler

Approximately 80 percent of poor people in Sub-Saharan Africa continue to depend on the agricultural sector for their livelihoods, but—unlike in other regions of the world—agriculture in Sub-Saharan Africa is characterized by very low yields due to agroecological features, poor access to services, lack of knowledge and inputs, and low levels of investment in infrastructure and irrigation. In addition, high population growth rates, especially in rural areas, intensify pressure on agricultural production and natural resources and further complicate the challenge of reducing poverty. Against this background, potential climate change poses a significant additional challenge to the future of agriculture in the region. Climate change could cause serious deterioration of rural livelihoods and increase food insecurity in Sub-Saharan Africa. Given these multiple challenges, the region's smallholders and pastoralists must adapt, in particular by adopting technologies to increase productivity and the stability and resilience of their production systems.

Improvements to agricultural productivity, often involving irrigation development, promote economic growth and provide a pathway out of poverty. Irrigation increases returns to poor households in terms of their physical, human, and social capital and enables smallholders to achieve higher yields and revenues from crop production. Irrigated farms also generate new employment opportunities through higher demand for farm labor. Poor consumers outside the agricultural sector also benefit from lower food prices because irrigation enables farmers to obtain more output per unit of input. Gains in agricultural productivity through irrigation can also stimulate national and international markets, improving economic growth. Nevertheless, rainfed farming continues to dominate agricultural production in Sub-Saharan Africa, covering around 97 percent of total cropland, which exposes agricultural production to high seasonal rainfall variability. Although irrigation systems have been promoted in the region, irrigation infrastructure has not expanded as expected mainly due to lack of demand for irrigated products, lack of access to affordable complementary inputs, poor market access, unfavorable topography, low-quality soils, and low incentives for agricultural intensification.

Despite these constraints, Sub-Saharan Africa has the potential to expand irrigation and increase agricultural productivity. A new generation of better designed irrigation projects and large, untapped water resources generate opportunities for irrigation investment in the region. In addition, with observed yields of less than one-third of the maximum attainable, the potential for productivity enhancement is significant. Finally, new investments in irrigation and improvements in agricultural productivity need complementary investments in roads, extension services, and access to markets.

This brief describes the results of a modeling exercise to forecast the economywide impacts of expanding irrigation and increasing agricultural productivity in Sub-Saharan Africa modeled under a relatively moderate SRES (Special Report on Emission Scenario) B2 climate change scenario, which was developed by the Intergovernmental Panel on Climate Change. The results were generated using a combination of economic models capable of assessing the security implications of a variety of development pathways under climate change for the period 2000–50.

THE IMPACT OF CLIMATE CHANGE ON AGRICULTURAL PRODUCTION, THE ECONOMY, AND WELFARE

The results of a scenario assuming no climate change (Scenario 1) project increased crop harvested area and crop production to the year 2050. Under this scenario, harvested area increases globally by about 3 percent between 2000 and 2050, which is equivalent to a total area of 1.35 billion hectares in 2050, 36 percent of which would be under irrigation. In Sub-Saharan Africa, irrigated area is projected to grow more than twice as fast as rainfed area over the same period, (79 compared with 34 percent), but the share of irrigated area is still only 4.5 percent in 2050 compared with 3.4 percent in 2000. Total agricultural production in Sub-Saharan Africa is projected to increase by 158 percent between 2000 and 2050, whereas the share of irrigated production is projected to increase from 9 to only 14 percent over the same 50-year period.

By comparison, a scenario of moderate climate change (Scenario 2) indicates that the world's crop harvested area and food production would decrease by 0.3 and 2.7 percent, respectively, by 2050. Under this second scenario, both irrigated area and irrigated production decrease, whereas rainfed production declines despite an increase in rainfed area. Hence, without proactive adaptation, climate change has a significant negative impact on agriculture. Both rainfed and irrigated harvested areas decrease in Sub-Saharan Africa under climate change, whereas rainfed production increases by 0.7 percent, and irrigated production drops by 15.3 percent. This sharp decrease in irrigated productivity occurs because some irrigated crops, such as wheat and sugarcane, are more susceptible to heat stress and reduced availability of water for irrigation. Under climate change, only 4.4 percent of the total crop harvested area is expected to be under irrigation by 2050, whereas irrigated production is expected to constitute 12.1 percent of total agricultural production in the region by 2050.

Climate change also has negative implications for gross domestic product (GDP) and child malnutrition. At the global level, GDP is expected to decrease by US$87 billion under a scenario of moderate climate change, which is equivalent to 0.09 percent of global GDP. For Sub-Saharan Africa, GDP losses due to climate change are estimated to be US$3.3 billion, equivalent to 0.2 percent of regional GDP. Furthermore, the number of malnourished children under five years old increases by almost 2 million in the region under this scenario.

In addition to the negative impacts described above, under a moderate climate change scenario without appropriate adaptation, the prices of agricultural commodities are projected to increase in both domestic and world markets. Real commodity prices for all cereals are projected to rise by 2050 due to increased land and water scarcity, as well as the impacts of climate change, biofuel development, increased population, and income- and growth-driven demand for food diversification. Meat prices are also expected to increase as a result of increased demand and higher animal feed prices.

STRATEGIES FOR ADAPTATION TO CLIMATE CHANGE

Adaptation by the agricultural sector can form a buffer against the negative impacts of climate change. Two different adaptation scenarios were used to evaluate the impact of adaptation on production and income. The first of these additional scenarios (Scenario 3) assumes expanded irrigated capacity, doubling the irrigated area in Sub-Saharan Africa, whereas the second of these adaptation scenarios (Scenario 4) considers productivity improvements in both rainfed and irrigated agriculture without expanded irrigated area, increasing Sub-Saharan African yields by 25 percent through investments in agricultural research and development and enhanced farm management practices.

Under Scenario 3, the expansion of irrigated area in the region from a very small base helps farmers achieve higher yields per hectare. This is followed by an increase in total crop production and a small decrease in the price of some agricultural commodities. Under this scenario, the expansion

Table 1 The impact of climate change on agriculture in Sub-Saharan Africa

Indicator	2000 baseline	Scenario 1: no climate change, 2050	Scenario 2: moderate climate change, 2050
Harvested area (thousand hectares)	181,618	246,363	244,585
Irrigated area (thousand hectares)	6,243	11,194	10,801
Rainfed area (thousand hectares)	175,375	235,169	233,784
Production (thousand metric tons)	484,199	1,250,491	1,231,158
Irrigated production (thousand metric tons)	43,398	175,561	148,701
Rainfed production (thousand metric tons)	440,800	1,074,930	1,082,457
Share of irrigated production (%)	9	14	12

SOURCE: IFPRI IMPACT model (2008).
NOTE: Scenario 2 is based on the Special Report on Emission Scenarios B2 scenario, developed by the Intergovernmental Panel on Climate Change. Results from the Hadley Global Circulation Model were used.

of irrigated area in Sub-Saharan Africa increases cereal production in the region by 5 percent, and meat production by 1 percent. No change was observed for root and tuber production. Even though Sub-Saharan Africa is not a key contributor to global food production, world food prices are projected to decline under the expanded irrigation adaptation scenario, which also leads to a small increase in GDP in Sub-Saharan Africa (0.007 percent, equivalent to US$113 million). This slight increase, however, is insufficient to counteract the regional GDP losses expected under climate change without proactive adaptation. Results show that doubling irrigated area also reduces the number of malnourished children under the age of five years by a small amount (0.3 million children).

Adaptation also has an impact on the factors of production, such as land, labor, and capital. Market prices for labor and capital increase as the economy expands. Changes in total crop production have a mixed effect on nonagricultural sectors, but the domestic and world prices of nonagricultural products increase under this scenario. An exception is food products, for which prices decline because production is promoted by higher supply and lower crop prices.

Under Scenario 4, improvements in agricultural productivity in both rainfed and irrigated agriculture enable farmers to obtain higher levels of output per unit of input and therefore total crop production increases (although the magnitude differs by crop type). Higher levels of agricultural productivity also result in a decline in production costs and consequently a decline in market prices. A 25 percent increase in agricultural productivity leads to significant reductions in domestic and world market prices for most agricultural commodities compared with projections without proactive adaptation (that is, Scenario 2). This adaptation scenario (Scenario 4) promotes GDP growth by 1.5 percent (US$26 billion), which more than offsets the initial reduction of 0.2 percent in GDP due to climate change as projected

Table 2 The impact of climate change and adaptation in Sub-Saharan Africa, 2050

Indicator	Scenario 1: no climate change, 2050	Scenario 2: moderate climate change, 2050 (without adaptation)[a]	Scenario 3: moderate climate change, 2050, with a doubling of irrigated area[b]	Scenario 4: moderate climate change, 2050, with a 25% increase in irrigated and rainfed crop yields[b]
Total production (thousand metric tons)	1,250,491	–1.5%	0.1%	18.0%
Rainfed production (thousand metric tons)	1,074,930	0.7%	–0.6%	17.9%
Irrigated production (thousand metric tons)	175,561	–15.3%	99.5%	23.4%
Total area (thousand hectares)	246,363	–0.7%	0.0%	0.0%
Rainfed area (thousand hectares)	235,169	–0.6%	–4.8%	0.0%
Irrigated area (thousand hectares)	11,194	–3.5%	100.0%	0.0%
Change in welfare (US$ millions)	--	1,786	119	15,435
Change in GDP				
US$ millions	--	–3,333	113	25,720
Percentage	--	–0.2%	0.0%	1.5%
Malnutrition (million children under age five)	30.2	32.0	31.7	30.4

SOURCE: IFPRI IMPACT and GTAP-W results (2008).
NOTE: Scenarios 2, 3, and 4 are based on the Special Report on Emission Scenarios B2 scenario, developed by the Intergovernmental Panel on Climate Change. Results from the Hadley Global Circulation Model were used.
[a] Indicates the percentage change with respect to Scenario 1.
[b] Indicates the percentage change with respect to Scenario 2.

under Scenario 2. Increased rainfed and irrigated crop productivity also significantly reduces the number of malnourished children (by 1.6 million children), essentially canceling out the increases in malnutrition caused by climate change.

In addition, market prices for rainfed land, irrigated land, and irrigation decline, while market prices for all other primary factors increase. The increase in the market price for unskilled labor is higher than for skilled labor under the second adaptation scenario (Scenario 4). Total production in nonagricultural sectors is also affected under Scenario 4. Reductions in total production are more pronounced for energy-intensive industry, other industry and services, and gas. Food products are positively affected, with production increasing by 1.4 percent. Domestic and world market prices increase for all nonagricultural sectors except for the food product industry.

IMPLICATIONS

Given the relatively low share of irrigated area in Sub-Saharan Africa, an increase in agricultural productivity achieves much larger benefits for the region than doubling irrigated area alone. The results show that improving crop yields in both rainfed and irrigated areas is a strategy that could almost completely offset the negative impact of climate change on productivity, GDP, prices, and child malnutrition. Substantial productivity gains are technically feasible in Sub-Saharan Africa because agricultural productivity is far below its potential.

An increase in agricultural productivity widely exceeds the GDP losses due to climate change, whereas increasing irrigated area alone does not offset GDP losses due to climate change. While these results are promising in terms of the potential to develop investment programs to counteract the adverse impacts of climate change, the climate change scenario implemented here is conservative in light of the range of potential climate change scenarios that could ultimately eventuate.

FOR FURTHER READING

Calzadilla, A., T. Zhu, K. Redhanz, R. S. J. Tol, and C. Ringler, *Economywide Impacts of Climate Change in Sub-Saharan Africa*, IFPRI Discussion Paper No. 873 (Washington, DC: International Food Policy Research Institute, 2009).

Alvaro Calzadilla (alvaro.calzadilla@zmaw.de) is a PhD student at the Unit Sustainability and Global Change, Hamburg University, and the Centre for Marine and Atmospheric Science, Hamburg, Germany. **Tingju Zhu** (t.zhu@cgiar.org) is a senior scientist in the Environment and Production Technology Division of the International Food Policy Research Institute (IFPRI), Washington, D.C. **Katrin Rehdanz** (katrin.rehdanz@ifw-kiel.de) is an assistant professor in the Department of Economics at Christian-Albrechts-University, Germany. **Richard S. J. Tol** (richard.tol@esri.ie) is a research professor at the Economic and Social Research Institute, Dublin, Ireland. **Claudia Ringler** (c.ringler@cgiar.org) is a senior research fellow in IFPRI's Environment and Production Technology Division.

This series of IFPRI Research Briefs is based on research supported by the Federal Ministry for Economic Cooperation and Development, Germany, under the project "Food and Water Security under Global Change: Developing Adaptive Capacity with a Focus on Rural Africa," which forms part of the CGIAR Challenge Program on Water and Food. Through collaboration with the Center for Environmental Economics and Policy in Africa, the Ethiopian Development Research Institute, the Ethiopian Economics Association, and the University of Hamburg, the project aims to provide policymakers and stakeholders in Ethiopia and South Africa with tools to better understand and analyze the consequences of global change—in particular climate change—and to form policy decisions that facilitate adaptation in these countries and beyond.

Financial Contributors and Partners

IFPRI's research, capacity strengthening, and communications work is made possible by its financial contributors and partners. IFPRI receives its principal funding from governments, private foundations, and international and regional organizations, most of which are members of the Consultative Group on International Agricultural Research (CGIAR). IFPRI gratefully acknowledges the generous unrestricted funding from Australia, Canada, China, Finland, France, Germany, India, Ireland, Italy, Japan, Netherlands, Norway, South Africa, Sweden, Switzerland, United Kingdom, United States, and World Bank.

Printed on alternative-fiber paper manufactured from agriculturally sustainable resources that are processed chlorine-free (PCF).

INTERNATIONAL FOOD POLICY RESEARCH INSTITUTE
sustainable solutions for ending hunger and poverty
IFPRI®
Supported by the CGIAR

EDRI
Ethiopia Development Research Institute

IFPRI Research Brief 15–16

HOW CAN AFRICAN AGRICULTURE ADAPT TO CLIMATE CHANGE? INSIGHTS FROM ETHIOPIA AND SOUTH AFRICA

Risk Aversion in Low-Income Countries

Experimental Evidence from Ethiopia

Mahmud Yesuf and Randy Bluffstone

Agricultural production in low-income developing countries is generally poorly diversified, focusing on rainfed staple crop production and raising livestock—activities that are inherently risky. Because of poorly developed or absent credit and insurance markets, it is difficult for farm households to pass any of these risks to a third party. Consequently, farm households base their investment and production decisions, in part, on the perceived risk of failure. A ramification of this is that households tend to be reluctant to adopt new agricultural technologies even when expected net returns are high. As such, a better understanding of risk behavior is essential for identifying appropriate farm-level strategies for adaptation to climate change by low-income farmers.

This brief is based on a study that estimates the magnitude and nature of farm household risk aversion in this context. Using an experimental approach, the authors examine the attitudes of Ethiopian smallholder farmers toward risk and suggest policy recommendations.

RISK IN AGRICULTURAL PRODUCTION IN DEVELOPING COUNTRIES

Households engaged in agricultural production in low-income countries face a number of risks, including crop yield risks due to variance in rainfall timing and level, animal mortality due to infectious livestock diseases, and changing output prices. Agricultural production is also affected by crop diseases, flooding, frost, illness of household members, war, and crime, all of which can have major effects on rural livelihoods.

The existence of such risks has been found to alter household behavior in ways that at first glance seem suboptimal. For example, it has been found that farm households use less fertilizer, fewer improved seed varieties, and lower levels of other production inputs than would have been the case had they simply maximized expected profits. Farmers' decisions to forgo welfare-improving opportunities because of perceptions of risk have significant policy implications. In rural areas of low-income countries, futures and insurance markets do not exist for most types of agricultural risk. Additionally, credit markets, which allow debtors and creditors to share risk, are thin. One policy response, therefore, is to develop or improve these markets by ensuring that insurance is available and by strengthening rural credit markets. Other measures could be to provide new technologies or inputs, together with long-term support through extension services.

Some advances have been made in these areas. Microcredit schemes abound in the developing world, allowing villagers to pursue production opportunities with less risk. Initiatives are also under way in Sub-Saharan Africa to develop crop insurance markets under the auspices of the World Bank and the World Food Programme.

While the existence of agricultural risk and its effects on low-income countries are well known, there are few empirical estimations of the magnitude and nature of household risk aversion in this context. Further, little is known about the basic household factors affecting risk behavior. Within low-income countries, there may be important linkages between risk aversion on the part of farm households and seemingly disparate elements such as household fertility, educational attainment, and gender dynamics. Working on these elements could thus improve outcomes for technology adoption.

To shed further light on this issue, the study uses an experimental approach to examine key determinants of risk aversion in smallholder agricultural production in Ethiopia. The experiment was administered as part of a 2002 survey of 1,522 households in 12 villages in the Ethiopian highlands of Amhara. The villages studied are typical of rural Ethiopia and representative of the nation as a whole. More than 70 percent of households sampled are illiterate, mean farm size is 1 hectare, and nearly all the households in the study rely heavily on agriculture to meet their consumption needs. The average household income in the sample is US$170, most of which is derived from agricultural production. Annual liquid cash availability is estimated at US$42, implying significant cash scarcity.

FACTORS THAT AFFECT RISK AVERSION

The study reveals a number of factors that affect households' reactions to risk when faced with new agricultural technologies. Some of these factors relate to the nature of the change in agricultural production, whereas others relate to households' past experiences and characteristics.

Results indicate that households are more sensitive to potential losses than they are to gains. Respondents who stand to lose as well as gain from adopting a new technology—even if the potential gain more than offsets the loss—are significantly more risk averse than those that face potential gains only. This finding strongly suggests agricultural extension efforts involving losses as well as gains may face systematic resistance by farmers in low-income, high-risk environments. However, once initial successes convince farmers that technologies are viable, risk aversion declines.

The study also identifies a positive relationship between the size of the expected payoff and the degree of risk aversion—that is, households are more risk averse the greater the expected return (even without the possibility of loss). Moreover, lower income households are more sensitive to risk than higher income households. Wealth—whether in the form of oxen, domestic animals, cash, or land—seems to reduce risk aversion.

In terms of past experiences, the study finds that farm households are more willing to accept risk if they have experienced successful past harvests. Similarly, households encountering a series of droughts may be more reluctant to undertake risky investment decisions.

Other factors that affect households' reaction to risk include household fertility (though not total household size), as well as the age and sex of the household head. The study suggests that families with a large number of dependents are more likely to avoid risky but potentially high-value technologies, such as improved seed varieties and chemical fertilizers. Furthermore, older household heads are more likely to avoid risk. Finally, male household heads—prevalent in Ethiopian farm households—were found to be less risk averse than female household heads.

POLICY IMPLICATIONS OF RISK ADVERSE BEHAVIOR OF FARMERS

In an effort to promote productivity-enhancing agricultural technologies in low-income countries like Ethiopia, the study makes several policy recommendations. First, the sensitivity of farm households to loss suggests that promoting technology with downside risks—even if potential gains are large—should be combined with insurance or other supporting measures. This support could be temporary, however. Once successes have convinced farmers that technologies are viable, risk aversion declines. Thus, the promotion of household-level technologies for adaptation to climate change must rely on proven methods that provide large gains and few losses, if any.

Second, the significant difference in risk-averting behavior between relatively poor and wealthy farm households suggests that as wealth accumulates, households are willing to take on more risk in exchange for higher returns. While early successes seem to be important, households should also be allowed to accumulate assets before they are challenged or tempted to take on more risky ventures. Further, the finding that households are more risk averse when the expected payoff is larger suggests that agricultural extension should start modestly before asking households to make larger changes.

In the longer run, of course, the development of private markets to spread risk is crucial. Indeed, broad-based economic development, including the development of credit and insurance markets, is the most certain way to reduce levels of risk aversion among farmers. Most practitioners would agree, however, that such developments are many years away, suggesting that interim risk mitigation solutions to promote rural development in low-income countries may be important for some time.

FOR FURTHER READING

Yesuf, M., and R. Bluffstone, *Risk Aversion in Low-Income Countries: Experimental Evidence from Ethiopia*, IFPRI Discussion Paper No. 715 (Washington, DC: International Food Policy Research Institute, 2007).

M. Yesuf (mahmudyesuf@yahoo.com) is a fellow at the Ethiopian Development Research Institute and Addis Ababa University. **R. Bluffstone** (bluffsto@pdx.edu) is an associate professor of economics at Portland State University.

This series of IFPRI Research Briefs is based on research supported by the Federal Ministry for Economic Cooperation and Development, Germany, under the project "Food and Water Security under Global Change: Developing Adaptive Capacity with a Focus on Rural Africa," which forms part of the CGIAR Challenge Program on Water and Food. Through collaboration with the Center for Environmental Economics and Policy in Africa, the Ethiopian Development Research Institute, the Ethiopian Economics Association, and the University of Hamburg, the project aims to provide policymakers and stakeholders in Ethiopia and South Africa with tools to better understand and analyze the consequences of global change—in particular climate change—and to form policy decisions that facilitate adaptation in these countries and beyond.

Financial Contributors and Partners

IFPRI's research, capacity strengthening, and communications work is made possible by its financial contributors and partners. IFPRI receives its principal funding from governments, private foundations, and international and regional organizations, most of which are members of the Consultative Group on International Agricultural Research (CGIAR). IFPRI gratefully acknowledges the generous unrestricted funding from Australia, Canada, China, Finland, France, Germany, India, Ireland, Italy, Japan, Netherlands, Norway, South Africa, Sweden, Switzerland, United Kingdom, United States, and World Bank.

Printed on alternative-fiber paper manufactured from agriculturally sustainable resources that are processed chlorine-free (PCF).

INTERNATIONAL FOOD POLICY RESEARCH INSTITUTE
sustainable solutions for ending hunger and poverty
IFPRI®
Supported by the CGIAR

EDRI
Ethiopia Development Research Institute

IFPRI Research Brief 15–17

HOW CAN AFRICAN AGRICULTURE ADAPT TO CLIMATE CHANGE? INSIGHTS FROM ETHIOPIA AND SOUTH AFRICA

Are Soil and Water Conservation Technologies a Buffer Against Production Risk in the Face of Climate Change?

Insights from Ethiopia's Nile Basin

Edward Kato, Claudia Ringler, Mahmud Yesuf, and Elizabeth Bryan

The agricultural sector in developing countries is particularly vulnerable to the adverse impacts of climate change. Given Ethiopia's dependence on agriculture and natural resources, any adverse agricultural effects will pose serious risks to economic growth and livelihoods across the country. Soil and water conservation technologies have been suggested as a key adaptation strategy for developing countries, particularly those in Sub-Saharan Africa, in light of increased water shortages, drought, desertification, and worsening soil conditions. According to a survey of 1,000 households in the Nile Basin of Ethiopia, more than 30 percent of farmers adopted soil and water conservation measures in response to perceived long-term changes in temperature and rainfall. Although soil and water conservation technologies are generally considered low-cost, they still engender risk for very low-income, risk-averse households, which are prevalent in rural Ethiopia. Thus, it is important to consider the impacts not only on crop yields, but also on risk levels.

This brief is based on a study that investigates the risk implications of various soil and water conservation technologies for crop production in Ethiopia's Nile River Basin. The analysis identifies technologies that increase and decrease crop production risk—with risk defined as the degree of yield variability—for the purpose of isolating which technologies are best suited to particular regions and agroecological zones. These results could be used to improve the geographical targeting of soil conservation techniques as part of efforts to promote farm-level adaptation to climate change.

THE USE OF SOIL AND WATER CONSERVATION TECHNOLOGIES IN ETHIOPIA

Soil and stone bunds are structures commonly built to control runoff and thus increase soil moisture and reduce soil erosion. However, constructing continuous bunds to protect broad tracts of land is costly and often difficult, so alternative methods of erosion control are also employed. These include grass strips and contour leveling, sometimes incorporating trees or hedgerows, to reduce runoff velocity and allow water to infiltrate and trap sediments. In addition, waterways help direct precipitation flows along specified pathways in farm fields; and water-harvesting structures, including dams, ponds, and diversions, ensure water availability in the dry season.

Using historical rainfall data from 1951 to 2000, Ethiopia's woredas (districts) were categorized as having either high or low rainfall. All of the woredas in Tigray and the Southern Nations, Nationalities and Peoples Region (SNNPR) fell into the low-rainfall category, whereas those in the Amhara, Oromiya, and Benishangul-Gumuz regions fell into a combination of both categories. Tigray appears to be the driest region of the five, and Oromiya had the highest average rainfall during 1951–2000.

Overall, the most commonly used soil and water conservation technologies by region are: soil bunds and stone bunds in Tigray, waterways and stone bunds in Amhara, soil bunds and waterways in Oromiya, waterways in Benishangul-Gumuz, and shade trees in SNNPR. Plots in low-rainfall areas have disproportionately more stone and soil bunds than plots in high-rainfall areas, and plots in high-rainfall areas have more waterways and irrigation. The results show clear spatial heterogeneity in the use of soil and water conservation technologies, suggesting that such technologies perform differently according to region and agroecological zone.

THE EFFECTS OF SOIL AND WATER CONSERVATION TECHNOLOGIES ON CROP YIELDS AND PRODUCTION RISK IN AREAS OF LOW AND HIGH RAINFALL

All soil and water conservation technologies considered in this study (that is, stone and soil bunds, grass strips, waterways, trees, and contours) show positive and highly significant impacts on crop output in the low-rainfall areas, but only waterways and trees show strong and significant positive effects in high-rainfall areas. Grass strips show the largest effect on crop yields among the technologies used in low-rainfall areas.

Although most of the technologies show significant positive effects on crop yields in low-rainfall areas, surprisingly, only soil bunds have a risk-reducing effect. This explains why almost 30 percent of the plots employ these techniques, and why other interventions that also have high positive impacts on yields are less common. In areas of high rainfall and high agricultural potential, most of the technologies considered in this study have risk-reducing effects, including soil and stone bunds, grass strips, waterways, and contours. Although both traditional and improved seeds increase average crop production in both low- and high-rainfall areas, they have different effects on production risk. Traditional seed is risk-reducing in both low- and high-rainfall areas, whereas improved seed

Table 1 Risk effects of soil and water conservation technologies on crop production by region and category of rainfall

	Tigray	Amhara		Oromiya		Benishangul-Gumuz		Southern Nations, Nationalities and Peoples Region
Technology employed	**Low rainfall**	**Low rainfall**	**High rainfall**	**Low rainfall**	**High rainfall**	**Low rainfall**	**High rainfall**	**Low rainfall**
Soil bunds		–		–	–		–	
Stone bunds			–		–			–
Grass strips			–		–	+	–	–
Waterway			–				–	–
Shade trees	+		–				–	–
Contour			–					
Traditional seed	–		–	–	–		–	
Improved seed			–		–			
Irrigation	+						–	
A combination of fertilizer, improved seed, and irrigation	–							

NOTE: A plus sign indicates significantly increased risk, whereas a minus sign indicates significantly reduced risk. Only results that are statistically significant at the 10 percent level are reported.

is only significantly risk-reducing in high-rainfall areas. These results suggest that, in attempting to adapt to climate change in low-rainfall areas, the choices of soil bunds and traditional seed are appropriate. Improved and traditional seed, stone and soil bunds, grass strips, waterways, and contours all appear to be promising adaptation strategies in high-rainfall areas.

THE EFFECTS OF SOIL AND WATER CONSERVATION TECHNOLOGIES ON CROP YIELDS AND PRODUCTION RISK BY REGION

The effects of soil and water conservation technologies within Ethiopia's Nile Basin vary not only by high- and low-rainfall area, but also by region (Table 1). The results for low-rainfall areas show that soil bunds are risk-reducing in Amhara and Oromiya, and that stone bunds are risk-reducing in the low-rainfall areas of SNNPR. Grass strips, waterways, and trees are only risk-reducing in SNNPR. Irrigation has no significant risk-reducing effect in any of the low-rainfall areas but shows a significant risk-increasing effect in the low-rainfall areas of Tigray. The risk-increasing aspect of irrigation in low-rainfall areas seems counterintuitive considering irrigation is intended to mitigate the adverse effects of low rainfall. Nevertheless, other studies have suggested reasons for failure of irrigation and water harvesting structures including poor technical design; lack of water, which could be stored in dry years; inappropriate and costly placement; and lack of community sensitization.

The results show that all technologies tend to reduce production risk in high-rainfall areas. Soil bunds are risk-reducing in Oromiya and Benishangul-Gumuz; stone bunds are risk-reducing in Amhara and Oromiya; grass strips are risk-reducing in Amhara, Oromiya, and Benishangul-Gumuz; waterways are risk-reducing in Amhara and Benishangul-Gumuz; trees are risk-reducing in Amhara and Benishangul-Gumuz; contours are risk-reducing in Amhara; and irrigation is risk-reducing in Benishangul-Gumuz."

CONCLUSIONS AND POLICY IMPLICATIONS

The results show that soil and water conservation technologies have significant impacts in reducing production risk in Ethiopia and could be part of the country's climate-proofing strategy. The results also show that one-size-fits-all recommendations are inappropriate given the differences in agroecologies and other factors. The performance of these technologies is location specific; therefore, programs aimed at promoting soil and water conservation measures as part of a climate change adaptation strategy should take these important differences into account.

FOR FURTHER READING

Kato, E., C. Ringler, M. Yesuf, and E. Bryan. 2009. *Soil and Water Conservation Technologies: A Buffer against Production Risk in the Face of Climate Change? Insights from the Nile Basin in Ethiopia.* Washington, DC: IFPRI Discussion Paper No. 871.

Edward Kato (e.kato@cgiar.org) and **Elizabeth Bryan** (e.bryan@cgiar.org) are research analysts and **Claudia Ringler** (c.ringler@cgiar.org) is a senior research fellow in the Environment and Production Technology Division (EPTD) of the International Food Policy Research Institute (IFPRI). **Mahmud Yesuf** (mahmudyesuf@yahoo.com) is a resident advisor and research fellow at the Kenya Institute for Public Policy Research and Analysis (KIPPRA), Nairobi, Kenya.

This series of IFPRI Research Briefs is based on research supported by the Federal Ministry for Economic Cooperation and Development, Germany, under the project "Food and Water Security under Global Change: Developing Adaptive Capacity with a Focus on Rural Africa," which forms part of the CGIAR Challenge Program on Water and Food. Through collaboration with the Center for Environmental Economics and Policy in Africa, the Ethiopian Development Research Institute, the Ethiopian Economics Association, and the University of Hamburg, the project aims to provide policymakers and stakeholders in Ethiopia and South Africa with tools to better understand and analyze the consequences of global change—in particular climate change—and to form policy decisions that facilitate adaptation in these countries and beyond.

Financial Contributors and Partners

IFPRI's research, capacity strengthening, and communications work is made possible by its financial contributors and partners. IFPRI receives its principal funding from governments, private foundations, and international and regional organizations, most of which are members of the Consultative Group on International Agricultural Research (CGIAR). IFPRI gratefully acknowledges the generous unrestricted funding from Australia, Canada, China, Finland, France, Germany, India, Ireland, Italy, Japan, Netherlands, Norway, South Africa, Sweden, Switzerland, United Kingdom, United States, and World Bank.

Printed on alternative-fiber paper manufactured from agriculturally sustainable resources that are processed chlorine-free (PCF).

INTERNATIONAL FOOD
POLICY RESEARCH INSTITUTE
sustainable solutions for ending hunger and poverty
IFPRI®
Supported by the CGIAR

IFPRI Research Brief 15–18

HOW CAN AFRICAN AGRICULTURE ADAPT TO CLIMATE CHANGE? INSIGHTS FROM ETHIOPIA AND SOUTH AFRICA

Green and Blue Water Accounting in the Limpopo and Nile Basins

Implications for Food and Agricultural Policy

Timothy B. Sulser, Claudia Ringler, Tingju Zhu, Siwa Msangi, Elizabeth Bryan, and Mark W. Rosegrant

Water scarcity is an increasingly critical issue for food production around the world. This is particularly true for the world's poorest region, Sub-Saharan Africa, due to its growing malnutrition and almost complete dependence on rainfed agriculture. Given that agriculture is the primary consumer of freshwater around the globe and the demand for domestic, industrial, and environmental water uses is steadily rising, strategies for the sustainable use of water in agriculture are urgently required.

Globally, most of the water used for food production is derived from precipitation or "green" water, and most of the water that reaches plants in irrigated systems also stems from precipitation. Irrigation water only dominates in dry seasons or in systems located in arid areas. At the same time, many production systems classified as rainfed involve small-scale applications of supplemental water intended to alleviate plant stress at critical stages of production (for example, via rainwater harvesting schemes). Thus, water applications for crop production follow a continuum, from purely rainfed to purely irrigated, but the majority of crops are produced from rainwater.

This brief is based on a paper that analyzes the changing contribution of both green and blue water for food production under various scenarios in the Limpopo and Nile river basins in Africa in order to inform the development of appropriate policy and investment responses.

BASELINE RESULTS

Under a "business-as-usual" or baseline scenario, agricultural water use in the Nile River Basin is expected to increase significantly, whereas agricultural water use development in the water-scarce Limpopo basin is very limited. In the Nile basin, irrigation water depletion is expected to grow by 2050, and crop water use from precipitation is projected to grow even more rapidly. As a result, irrigation is projected to account for a declining share in total crop water consumption despite a projected expansion of irrigated area. In the Limpopo basin in southern Africa, on the other hand, increased non-irrigation water demand and a relative lack of opportunities to further develop water sources are projected to cause a decline in irrigation water growth. Rainfed crop production is also expected to contract, while water productivity is expected to increase; total crop area in the basin is expected to remain roughly constant by 2050.

Water productivity for crops—defined as the amount of harvested commodity per unit of consumptive water used—is a key metric of water-use efficiency for different agricultural production systems around the world. Current water-productivity levels in the Limpopo and Nile basins are below the global average. In both African basins, however, water productivity is projected to increase over time because of technological change and growing competition with other uses.

Although food security is projected to improve nominally in the countries of the Limpopo and Nile river basins by 2050, projected changes in calorie availability and the number of malnourished children indicate that these improvements are not likely to occur until after 2025 in most countries. Only a handful of the 10 countries of the Nile River Basin are projected to achieve the 2015 hunger reduction target of the Millennium Development Goals by the year 2050.

SCENARIOS OF INVESTMENT IN AGRICULTURAL TECHNOLOGY AND R&D

The outcomes of three scenarios of investment in agricultural technology and research and development (R&D) were compared with those of the baseline scenario to shed light on possible policy interventions to improve food and water security in the Limpopo and Nile basins. Scenario A, a low-investment scenario, presents a fairly pessimistic view of future agricultural production and includes a further reduction in the already slowing rates of investment in agricultural R&D. Scenario B, a higher investment scenario, presents an optimistic outlook, assuming that policymakers will prioritize investment in agricultural productivity, particularly in the developing world. Scenario C supposes even greater investment in yield improvements and the intensification of existing agricultural systems. Under this scenario, agricultural productivity investments are further enhanced with investments in irrigation infrastructure and other critical areas that promote reduced poverty and malnutrition.

Food and water security outcomes differ markedly across the future scenarios, indicating the importance of appropriate policy choices and investments. Under Scenario A, calorie availability is strongly reduced such that the number of malnourished children significantly increases, both in absolute numbers and in terms of prevalence. Under this scenario, the number of malnourished children in the countries of the Limpopo and Nile basins increases by 0.8 and 7.7 million compared with baseline levels, respectively, in 2050 (8.5 million in total). Under scenarios B and C, on the other hand, the number of malnourished children decreases by 5.8 and 13.2 million, respectively, due to per capita calorie availability increases across the Limpopo and Nile basin countries averaging 800 kilocalories under Scenario B and 1,600 kilocalories under Scenario C.

In the Limpopo River Basin, slowing yield growth under Scenario A leads to increases in both irrigated and rainfed harvested areas to meet future food demand (Table 1). Consequently, for all crops except rice, consumptive water use increases to 2050 compared with baseline levels. In contrast, the productivity increases under scenarios B and C require less expansion in irrigated and rainfed areas and consumptive water use compared with the baseline scenario. The expanded irrigation infrastructure posited under Scenario C displaces some of the rainfed area compared with the other scenarios, but total consumptive water use remains below the levels forecast under the other scenarios.

In the Nile River Basin, the patterns of total production, area, and water consumption across the various scenarios are similar for the major crops, but area expansion for both irrigated and rainfed production under Scenario A is driven higher than baseline levels due to the suppression of yield growth. Water consumption increases due to the expanded area, and production levels remain below those forecast under the baseline scenario. Scenarios B and C result in higher output levels on less land and with lower water consumption. Compared with the Limpopo basin, changes in water productivity are much lower in the Nile basin. Under Scenario C, water productivity declines for irrigated cereals as lower food prices and resulting higher food demand induce rapid crop area expansion that cannot be fully met with the available resources.

Table 1 Projected percentage change in consumptive water-use productivity for cereals compared with baseline levels, 2025 and 2050

Region	Water source	Scenario A: Pessimistic/ low investment 2025	Scenario A 2050	Scenario B: Optimistic/ high investment 2025	Scenario B 2050	Scenario C: Very optimistic/ very high investment 2025	Scenario C 2050
Limpopo basin	Irrigated cereals	–12.7	–12.9	13.9	19.2	20.6	18.2
	Rainfed cereals	–12.7	–10.6	14.2	17.6	27.3	27.5
Nile basin	Irrigated cereals	–2.7	–3.2	2.4	8.1	–1.5	5.5
	Rainfed cereals	–11.4	–17.2	12.6	25.8	17.7	38.3
Global	Irrigated cereals	–6.0	–9.3	6.0	14.1	5.7	17.3
	Rainfed cereals	–5.6	–8.5	5.8	14.4	7.5	20.8

NOTE: Projections are based on kilograms of harvested cereal per cubic meter of consumptive water use.

POLICY IMPLICATIONS AND CONCLUSION

The results of this study emphasize the importance of more disaggregated analyses, given opposing trends for consumptive water use in the Limpopo and Nile river basins. In the Limpopo River Basin, growing water scarcity suggests the need for investment in technologies that enhance irrigated and rainfed crop yields. In the Nile River Basin, on the other hand, irrigated crop productivities are fairly high and are achieved with little complementary precipitation (chiefly in Egypt), so the focus needs to be on expanding irrigated areas elsewhere and improving rainfed crop productivity.

Based on our analysis, we find that an approach that combines strategies targeting both blue and green water management with other complementary investments in rural agricultural development has the potential to improve the lives of many poor people.

FOR FURTHER READING

Sulser, T. B., C. Ringler, T. Zhu, S. Msangi, E. Bryan, and M.W. Rosegrant; *Green and Blue Water Accounting in the Limpopo and Nile Basins: Implications for Food and Agricultural Policy.* IFPRI Discussion Paper No. 907 (Washington, DC: International Food Policy Research Institute, 2009).

Timothy B. Sulser (t.sulser@cgiar.org) is a scientist; **Claudia Ringler** (c.ringler@cgiar.org) and **Siwa Msangi** (s.msangi@cgiar.org), senior research fellows; **Tingju Zhu** (t.zhu@cgiar.org), a senior scientist; and **Elizabeth Bryan** (e.bryan@cgiar.org), a research analyst in the Environment and Production Technology Division (EPTD) of the International Food Policy Research Institute. **Mark W. Rosegrant** (m.rosegrant@cgiar.org) is EPTD's division director.

This series of IFPRI Research Briefs is based on research supported by the Federal Ministry for Economic Cooperation and Development, Germany, under the project "Food and Water Security under Global Change: Developing Adaptive Capacity with a Focus on Rural Africa," which forms part of the CGIAR Challenge Program on Water and Food. Through collaboration with the Center for Environmental Economics and Policy in Africa, the Ethiopian Development Research Institute, the Ethiopian Economics Association, and the University of Hamburg, the project aims to provide policymakers and stakeholders in Ethiopia and South Africa with tools to better understand and analyze the consequences of global change—in particular climate change—and to form policy decisions that facilitate adaptation in these countries and beyond.

Financial Contributors and Partners

IFPRI's research, capacity strengthening, and communications work is made possible by its financial contributors and partners. IFPRI receives its principal funding from governments, private foundations, and international and regional organizations, most of which are members of the Consultative Group on International Agricultural Research (CGIAR). IFPRI gratefully acknowledges the generous unrestricted funding from Australia, Canada, China, Denmark, Finland, France, Germany, India, Ireland, Italy, Japan, the Netherlands, Norway, the Philippines, South Africa, Sweden, Switzerland, the United Kingdom, the United States, and the World Bank.

Printed on alternative-fiber paper manufactured from agriculturally sustainable resources that are processed chlorine-free (PCF).

INTERNATIONAL FOOD POLICY RESEARCH INSTITUTE
sustainable solutions for ending hunger and poverty
IFPRI®
Supported by the CGIAR

IFPRI Research Brief 15–19

HOW CAN AFRICAN AGRICULTURE ADAPT TO CLIMATE CHANGE? INSIGHTS FROM ETHIOPIA AND SOUTH AFRICA

Climate Change Impacts in Ethiopia
Hydro-Economic Modeling Projections

Gene J.-Y. You and Claudia Ringler

Two factors critical to assuring food security, whether at the local or the global level, are increasing crop productivity and increasing access to sustainable water supplies. These factors are also vital to the economic success of agriculture, which is particularly important in Ethiopia given that the sector accounts for about 41 percent of the country's gross domestic product (GDP), produces 80 percent of its exports, employs 80 percent of the labor force, and is a major source of income and subsistence for the nation's poor.

Extreme hydrological variability and seasonality have constrained Ethiopia's past economic development by negatively affecting crop production—chiefly through droughts—and by destroying roads and other infrastructure due to flooding. As climate change unfolds, average climatic variables will shift, and weather variability will intensify, exposing Ethiopian agriculture to higher levels of risk and jeopardizing economic growth, food security, and poverty reduction. Most of the studies that have helped clarify the strong relationship between climate variability and Ethiopia's GDP have been based on historical data, but future projections of climatic changes are critical to our understanding of the evolution of hydrological conditions in Ethiopia and our ability to extrapolate their associated effects on economic growth.

This brief is based on a paper that models the effects of three climate change–driven factors on the Ethiopian agricultural sector and overall economy. The model investigates the economic impact of water constraints on rainfed food production, changes in CO_2 fertilization due to increased atmospheric CO_2, and losses due to floods.

OVERVIEW OF THE MODEL

The study assesses selected global circulation models from the Fourth Assessment Report of the Intergovernmental Panel on Climate Change (2007) to project changes in water stress and flood events to 2050, together with changes in CO_2 concentration. The projections are compared with a 1990–2000 baseline period for three different emission scenarios. Outputs from the projections are then translated into impacts on crop yield due to water constraints, flood damage, and fertilization effects. The study then uses a multimarket model simulating the period 2003–50 to analyze the effects of changes in water constraints, flood damage, and fertilization on economic indicators such as agricultural GDP growth, overall GDP growth, and the poverty rate.

In order to compare projections with baseline conditions in the absence of climate change, the model incorporates new economic parameters for projected population growth and baseline economic growth to reflect recently updated observations. Importantly, the study examines the potential of the irrigation expansion recently proposed by the Ethiopian Ministry of Water Resources to buffer the effects of climate change.

NEGATIVE EFFECT ON GDP GROWTH

Although CO_2 can increase vegetative growth and biomass, a number of other factors like fertilizer availability must be in place for this potential to be realized. Results using the multimarket model show that CO_2 fertilization is expected to increase the rate of agricultural GDP growth and that this may be further enhanced by the Ethiopian Ministry of Water Resources' proposed irrigation development.

Projections from the global circulation modeling show increased annual mean rainfall and an increase in evapotranspiration to the year 2050, although the magnitude of the variability in these parameters is larger than the change in mean values. The incremental variability of precipitation, which translates as fluctuating rainfall, reduces the availability of a stable water supply and increases the risk of floods. The frequency of low-probability extreme events is expected to increase as well. When these projections are translated into impacts on crop yields due to water constraints and flood damage, results from multimarket modeling indicate that flood damage—mainly influenced by weather variability rather than changes in the means—has a larger depressing effect on agricultural and overall GDP growth.

These results demonstrate that the negative impact on GDP growth actually stems from hydrological variability rather than water supply constraints. When the effects of all three climate factors are evaluated together, flood damage still drives the overall impacts. The difference between the three climate scenarios is minimal, with the worst conditions occurring for the scenario that embodies the most extreme climatic changes (Scenario A2). Implementation of the proposed irrigation development has a positive, but limited, buffering effect on agriculture GDP growth under climate change conditions for all three scenarios. The improvement is visible, but it does not change the main adverse impact of climate change on growth (Figure 1).

Analysis of climate change impacts on five agriculture subsectors that are important for smallholder subsistence farmers and pastoralists shows that climate change increases poverty, even with increased irrigation development.

POLICY IMPLICATIONS

In Ethiopia, climate change is expected to intensify the already high hydrological variability and frequency of extreme events. More than changes in mean annual rainfall, these two factors may have a significant negative effect on the development of the agricultural sector and on the Ethiopian economy as a whole. Droughts impair agricultural productivity and may lock subsistence farmers into poverty traps, whereas recurrent flooding can have long-term negative effects on agricultural GDP by directly damaging crops and by destroying roads, thereby exacerbating the inadequacy of transport infrastructure and consequently limiting access to markets. Omitting climate change impacts from future investment analyses will lead to suboptimal investment decisions. Our analysis can therefore support decisionmaking by identifying development strategies that offer the highest resilience to future climate change.

Based on the analysis, it appears that investments in multipurpose water infrastructure, such as reservoirs, detention ponds, and small dams, have a high potential to address increased hydrologic variability by increasing water storage and regulating water flows, while at the same time providing water for irrigation. The benefit of the Ethiopian government's proposed irrigation development could thus be further enhanced through a focus on multipurpose storage infrastructure in high-risk flood areas.

Over time, the nonagricultural sector will increasingly dominate Ethiopia's GDP growth, but most of the population will continue to depend on agriculture for survival. Given important concerns about the country's wealth distribution and poverty levels—and given that agriculture is highly vulnerable to climate variability and change—Ethiopia should start to invest in agriculture and rural water today.

FOR FURTHER READING

You, G. J.-Y., and C. Ringler. 2010. *Hydro-Economic Modeling of Climate Change Impacts in Ethiopia*. IFPRI Discussion Paper No. 960. (Washington, DC: International Food Policy Research Institute, 2010).

Figure 1 Simulations of GDP growth based on changes in water constraints, flood damage, and fertilization

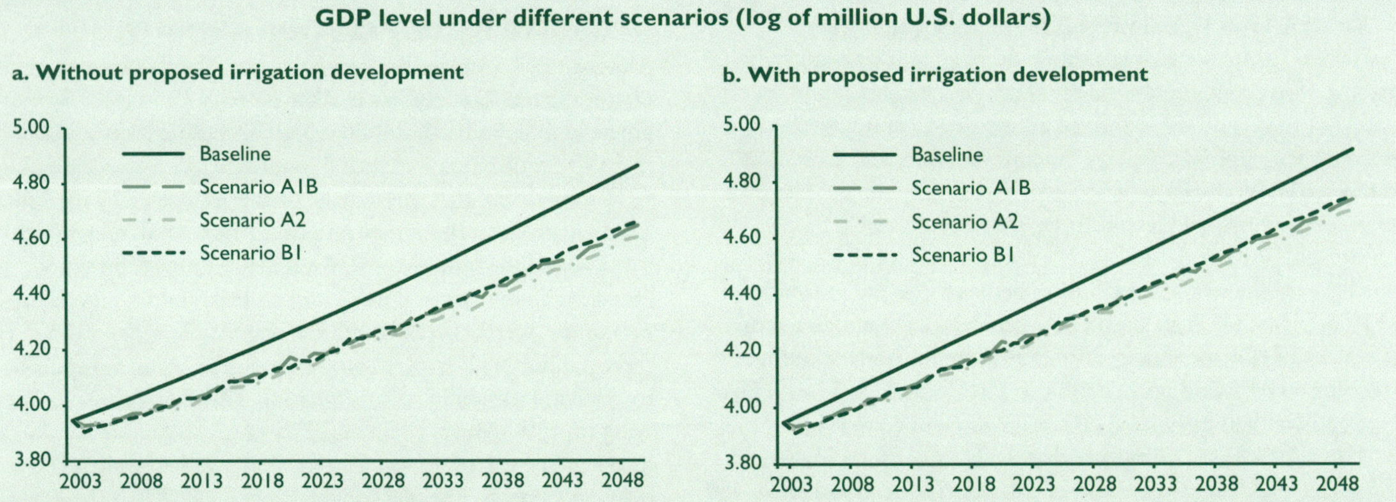

NOTE: The baseline scenario provides GDP projections without climate change. The worst effect on GDP growth is observed under Scenario A2, which corresponds to the most extreme climatic changes.

Gene J.-Y. You (genejyu@ntu.edu.tw) is an assistant professor in the Department of Civil Engineering at National Taiwan University. **Claudia Ringler** (c.ringler@cgiar.org) is a senior research fellow in the Environment and Production Technology Division of the International Food Policy Research Institute.

This series of IFPRI Research Briefs is based on research supported by the Federal Ministry for Economic Cooperation and Development, Germany, under the project "Food and Water Security under Global Change: Developing Adaptive Capacity with a Focus on Rural Africa," which forms part of the CGIAR Challenge Program on Water and Food. Through collaboration with the Center for Environmental Economics and Policy in Africa, the Ethiopian Development Research Institute, the Ethiopian Economics Association, and the University of Hamburg, the project aims to provide policymakers and stakeholders in Ethiopia and South Africa with tools to better understand and analyze the consequences of global change—in particular climate change—and to form policy decisions that facilitate adaptation in these countries and beyond.

Financial Contributors and Partners

IFPRI's research, capacity strengthening, and communications work is made possible by its financial contributors and partners. IFPRI receives its principal funding from governments, private foundations, and international and regional organizations, most of which are members of the Consultative Group on International Agricultural Research (CGIAR). IFPRI gratefully acknowledges the generous unrestricted funding from Australia, Canada, China, Denmark, Finland, France, Germany, India, Ireland, Italy, Japan, the Netherlands, Norway, the Philippines, South Africa, Sweden, Switzerland, the United Kingdom, the United States, and the World Bank.

Printed on alternative-fiber paper manufactured from agriculturally sustainable resources that are processed chlorine-free (PCF).

INTERNATIONAL FOOD
POLICY RESEARCH INSTITUTE
sustainable solutions for ending hunger and poverty
IFPRI®
Supported by the CGIAR

IFPRI Research Brief 15-20

HOW CAN AFRICAN AGRICULTURE ADAPT TO CLIMATE CHANGE? INSIGHTS FROM ETHIOPIA AND SOUTH AFRICA

Climate Change Impacts on Food Security in Sub-Saharan Africa

Insights from Comprehensive Climate Change Modeling

Claudia Ringler, Tingju Zhu, Ximing Cai, Jawoo Koo, and Dingbao Wang

According to the Fourth Assessment Report of the Intergovernmental Panel on Climate Change, warming in Sub-Saharan Africa (SSA) is expected to be greater than the global average, and rainfall will decline in certain areas. Global circulation models (GCMs), which provide an understanding of climate and project climate change, tend to agree that temperatures are increasing across the region, but models vary widely regarding predicted changes in precipitation—with the exception of some agreement that precipitation decreases from June to August in southern Africa and increases from December to February in eastern Africa. Whether the Sahel will be more or less wet in the future remains uncertain. Given the limited agreement of GCMs, the University of Illinois and the International Food Policy Research Institute (IFPRI) developed a global comprehensive climate change scenario combining 17 models selected for their past performance in predicting temperature and precipitation.

This brief is based on a study that integrates these esults with a process-based crop simulation model and IFPRI's International Model for Policy Analysis of Agricultural Commodities and Trade (IMPACT) to assess climate change outcomes for SSA. The modeling approach employed for the study considered three possible climate change impacts on crop production: (1) direct effects on rainfed yields th ough changes in temperature and precipitation, (2) indirect effects on irrigated yields f om changes in temperature and available irrigation water (including precipitation), and (3) autonomous adjustments to area and yield due to price effects and changes in trade flows. Overall, results indicate that climate change impacts, as evidenced by declining crop yields, are less severe in SSA compared with other regions like Asia because yield levels are much lower to start with, and fertilizer application is limited. These same conditions make SSA much more vulnerable to climate change, particularly because low yield levels and limited agricultural inputs are combined with a high dependence on rainfed agriculture and high poverty levels.

Figure 1 Projected changes in Sub-Saharan African crop yields due to climate change, 2050

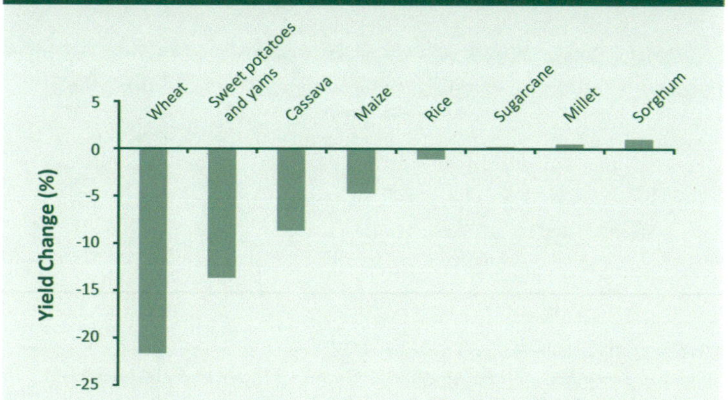

Source: Compiled by authors based on IFPRI IMPACT modeling projections.

THE IMPACT OF CLIMATE CHANGE ON FOOD OUTCOMES

Cereal production growth for a range of crops in SSA is projected to decline by a net 3.2 percent in 2050 as a result of climate change. Increased area expansion of 2.1 percent partially compensates for an overall yield growth decline of 4.6 percent. The largest negati e yield impacts are projected for wheat—of which the region grows very little—followed by sweet potatoes. Overall, millet and sorghum yields are projected to be slightly higher under climate change, probably given their higher tolerance to higher temperatures and drought stress (see Figure 1).

World prices are a key indicator of the effects of climate change on agriculture and, even more importantly, on food affo dability and security. Food prices increase for all staple crops because climate change acts as an additional stressor on the already tightening price outlook. Under climate change, maize, rice, and wheat prices in 2050 are projected to be 4, 7, and 15 percent higher than under the historic climate scenario. Moreover, prices of other important crops in the region also increase—for sweet potatoes and yams by 26 percent, cassava by 20 percent, millet by 5 percent, and sorghum by 4 percent.

Higher food prices are projected to dampen demand for food, as the affordability of nearly all agricultural commodities, including basic staples and livestock products, declines under climate change. As a result, per capita calorie availability across SSA is projected to decline by 1.3 percent, or 37 kilocalories per capita per day. While this change appears comparatively small, distributional effects are likely to be significant, and those who can least afford to reduce caloric intake are likely to be hit the hardest. The largest drop in calorie availability, at 2.6 percent, is projected for the central zone, which already had the lowest per capita calorie availability to begin with. Under these conditions, on average, the central zone would be close to the minimum per capita daily calorie availability of 2,000 kilocalories recommended for a healthy and productive life.

Climate change increases the number of malnourished children in both 2030 and 2050. Without climate change, child malnutrition levels in SSA are projected to decline from 28 percent in 2000 to 24 percent in 2030 and 19 percent in 2050, while the absolute number of malnourished children would still increase from 30 million children in 2000, to 38 million in 2030, before reverting to 30 million in 2050, given the continued rapid increase in population growth in the region. Under climate change, child malnutrition would increase by an additional 0.5 million children in 2010, would be higher by 1 million children in 2030, and would still be higher by 0.6 million children by 2050.

Changes in agricultural trade fl ws as a result of climate change are driven by changes in the local biophysical and socioeconomic environment, as well as a wide-ranging set of local, regional, national, and international trade policies. Across SSA, little change in net cereal imports is expected as a result of climate change because small increases and decreases in net cereal imports of particular countries effecti ely balance each other out. At the subregional level, eastern Africa is projected to experience the largest increase in net cereal imports due to climate change (15 percent) as a result of declining maize yields. For the Sudano–Sahelian zone, a steep decline in net cereal imports is also projected (6 percent), again driven by local changes in maize yields.

POLICY CONCLUSIONS

Even without climate change, SSA remains the most food-deprived region worldwide and the only one with projected increases in childhood malnutrition over the next two decades despite recent increases in economic prosperity and gross domestic product, which were generated through agriculture. Compared with historic climate records, climate change will cause shifts in yield and area growth and increased food prices, thereby lowering food affo dability, reducing calorie availability, and increasing childhood malnutrition. Cereal production growth in the region is projected to decline by 3.2 percent as a result of climate change, with increased area expansion of 2.1 percent partially compensating declines in yield growth of 4.6 percent. The most potent fo ce for reducing malnutrition—particularly in SSA—is raising food availability and rural incomes through increased agricultural productivity. Agricultural productivity enhancements will thus be critical in counteracting the adverse impacts of climate change in the region.

FOR FURTHER READING

Ringler, C., T. Zhu, X. Cai, J. Koo, and D. Wang. 2010. *Climate Change Impacts on Food Security in Sub-Saharan Africa: Insights from Comprehensive Climate Change Scenarios.* IFPRI Discussion Paper No. 1042. (Washington, DC: International Food Policy Research Institute, 2010).

Claudia Ringler (c.ringler@cgiar.org) is a senior research fellow, **Tingju Zhu** (t.zhu@cgiar.org) a senior scientist, and **Jawoo Koo** (j.koo@cgiar.org) a research fellow in the Environment and Production Technology Division of the International Food Policy Research Institute, Washington, DC. **Ximing Cai** (xmcai@illinois.edu) is an associate professor in the Department of Civil and Environmental Engineering at the University of Illinois, Urbana-Champaign, and, at the time this study was conducted, **Dingbao Wang** (diwang@mail.ucf.edu) was a PhD student at the University of Illinois, Urbana-Champaign.

This series of IFPRI Research Briefs is based on research supported by the Federal Ministry for Economic Cooperation and Development, Germany, under the project "Food and Water Security under Global Change: Developing Adaptive Capacity with a Focus on Rural Africa," which forms part of the CGIAR Challenge Program on Water and Food. Through collaboration with the Center for Environmental Economics and Policy in Africa, the Ethiopian Development Research Institute, the Ethiopian Economics Association, and the University of Hamburg, the project aims to provide policymakers and stakeholders in Ethiopia and South Africa with tools to better understand and analyze the consequences of global change—in particular climate change—and to form policy decisions that facilitate adaptation in these countries and beyond.

Financial Contributors and Partners

IFPRI's research, capacity strengthening, and communications work is made possible by its financial contributors and partners. IFPRI receives its principal funding from governments, private foundations, and international and regional organizations, most of which are members of the Consultative Group on International Agricultural Research (CGIAR). IFPRI gratefully acknowledges the generous unrestricted funding from Australia, Canada, China, Denmark, Finland, France, Germany, India, Ireland, Italy, Japan, the Netherlands, Norway, the Philippines, South Africa, Sweden, Switzerland, the United Kingdom, the United States, and the World Bank.

Printed on alternative-fiber paper manufactured from agriculturally sustainable resources that are processed chlorine-free (PCF).

Made in United States
Troutdale, OR
11/03/2026

56844434R00029